Why isn't the first page of a book Page 1?
Why must you turn through page after page of publisher's froth before getting down to the business at hand?
I've always wanted to call this page "Page 1" but, until now, the world has intervened, insisting that this first sheet be left blank, or, if space were at a premium, that it be used as a title page. Why a title should be repeated so close behind the cover is something that has never been made clear to me, but, just in case you've already forgotten, the book you now hold in your hands is called Passion's Playmate. Now let's get on with the story.

Here you are on page 1, about to see what I've been doing for the past 26 years. My underground career has produced over 400 designs but fewer than one in twenty were ever built. My actually-built-to-number-designed ratio is dismal. It's a record matched only by the equally dismal records of many other architects who learned, as I did, that in this profession the bright bubbles of new design commissions usually disappear into thin air. In the engineering profession, "back to the drawing board" means we built it and it didn't work, but in architecture it means go dream another dream; the project has just been cancelled.

So this book can very well be seen as a catalog of failures. Maybe that's just what it is. If it has any value, its usefulness may lie in proving that there are as many ways to design an

She asked me if I'd dedicate this book to her but I'm indebted to too many people to do that. I can offer her only 59/64ths of this front page: To my step-daughter, Kirsten Elizabeth Kelly, from the water miser and essay checker.

earth-sheltered building as there are to design the other kind. The other kind, of course, is what we see all around us: embarrassingly naked and wasteful buildings.

Another value of what follows is that it portrays a winner. Underground architecture, in spite of my personal record with it, is bound to succeed. It has the most powerful of all possible allies: the living world. Rather than stand alone in contrast to Nature, underground architecture lies in her arms.

The ideal – the goal – of a truly appropriate architecture – that is, invisibility – has hardly even been approached in this book. There's still a long way to go. Years, perhaps decades, may pass before we learn to build as if the land were sacred. Still, every other creature builds that way. And the land is sacred. We know, now, how to build without condemning building sites to the unending torture of life under asphalt and concrete, and so, inching our way forward, we can hope for the best.

What follows is roughly chronological, from my underground awakening in 1964 (after 11 years of doing nothing but surface architecture) to this year, 1990.

MALCOLM WELLS, underground on Cape Cod.

Next, I want to dedicate this page in its entirety, with love, to my wife, one Karen North Wells, whose paintings will be in a book of their own someday. She found all the typos in this book except for the ones you'll no doubt discover.

Now for other business ...

 But if you write to me and ask to use any of the designs that follow I will, with a few exceptions (those being the jobs currently active for specific clients) give you permission.

Ordering information:

- Retail price: $14.95 (add sales tax if you are in Massachusetts).
- Mail order price: $14.95, postpaid (book rate) within the U.S. To Canada, send $18 in U.S. funds. (See "Foreign Payments.")
- First Class (priority mail): $17, postpaid, in the U.S.
- Overseas airmail price: $22, postpaid. (See "Foreign Payments" below.)
- Foreign payments: must be in U.S. dollars, on international money orders, or checks drawn on New York banks.
- Wholesale price: 40% discount on orders of 10 or more copies, plus shipping cost.

I.S.B.N. 0-9621878-1-X

Printed by Raven Rocks Press
Beallsville, Ohio

▶ My address:
673 Satucket Road
Brewster, MA 02631, U.S.A.
(508) 896 6850

Printed on recycled paper.

When I was just a little squirt, growing up in New Jersey, I saw, every Sunday, in the town of Cedarbrook, a strange concrete structure. It was just one of the many things I came to know on the way to my grandparents' house. Not till I was grown up, and an architect, did it occur to me that that strange structure was anything really special. How had a Wrightian-cantilevered-reinforced-concrete pedestal come to a tiny village in a farm community? How had plants appeared on top of it? How did they survive? Who built it? And when?

I don't know if the structure is still there. I don't want to find out. It's very likely to have been swept away, along with the town itself, in the tidal wave of suburban sprawl around Philadelphia.

Somewhere between the two, between the suburban sprawl and the graceful plant-covered structure, the idea of a gentler kind of architecture found its way into my head, but it took its own sweet time in getting there. I had been a practicing architect for 11 years before I began to wonder if there was something less destructive than asphalt and concrete and toxic green lawns.

Marguerite Kinney Wells got started in '95 and is still going strong. She's always been my favorite mother, so, as a token of my love, I hereby dedicate almost this whole page just to her. Ma, can you read all this tiny print?

The answer, of course, was all around me. The natural world teemed with examples. That the surface of the earth was meant for living plants, instead of dead buildings and asphalt, never got through to me until the environmental consciousness of the sixties unclogged my head. That's when the earth-cast buildings of Riviera architect Jacques Couelle, the strange pedestal in New Jersey, and Frank Lloyd Wright's underground theatre got together and decided it was time for me to wake up.

In 1959, at Taliesin West, I stepped out of the hot desert sunlight into this little open-to-the-air theatre, and marveled for a moment or two at Mr. Wright's genius, his ability to carry a design through into the tiniest of details, before it struck me that I was suddenly cool and comfortable there under a mantle of earth.

It took me only 5 more years to get the message. In 1964, I suddenly had a brilliant and original idea: buildings should be underground!

The revelation was so compelling I knew that all I needed to do was let the world know about it and a new architecture would be born.

Within days, I had designed underground cities, underground highways, underground shopping centers, and this house, all of which the editors of Progressive Architecture, perhaps feeling the mounting pressure of the times, agreed to publish.

In the February issue of 1965, the drawings appeared along with my polemic against everything that had ever been built on the surface of the earth. It was now only a matter of time before all new buildings would be built this way. I braced myself for the world's reaction.

The world, it appeared, had other fish to fry. A hundred letters and a few lecture dates were all that came of it — except for the reaction to this house.

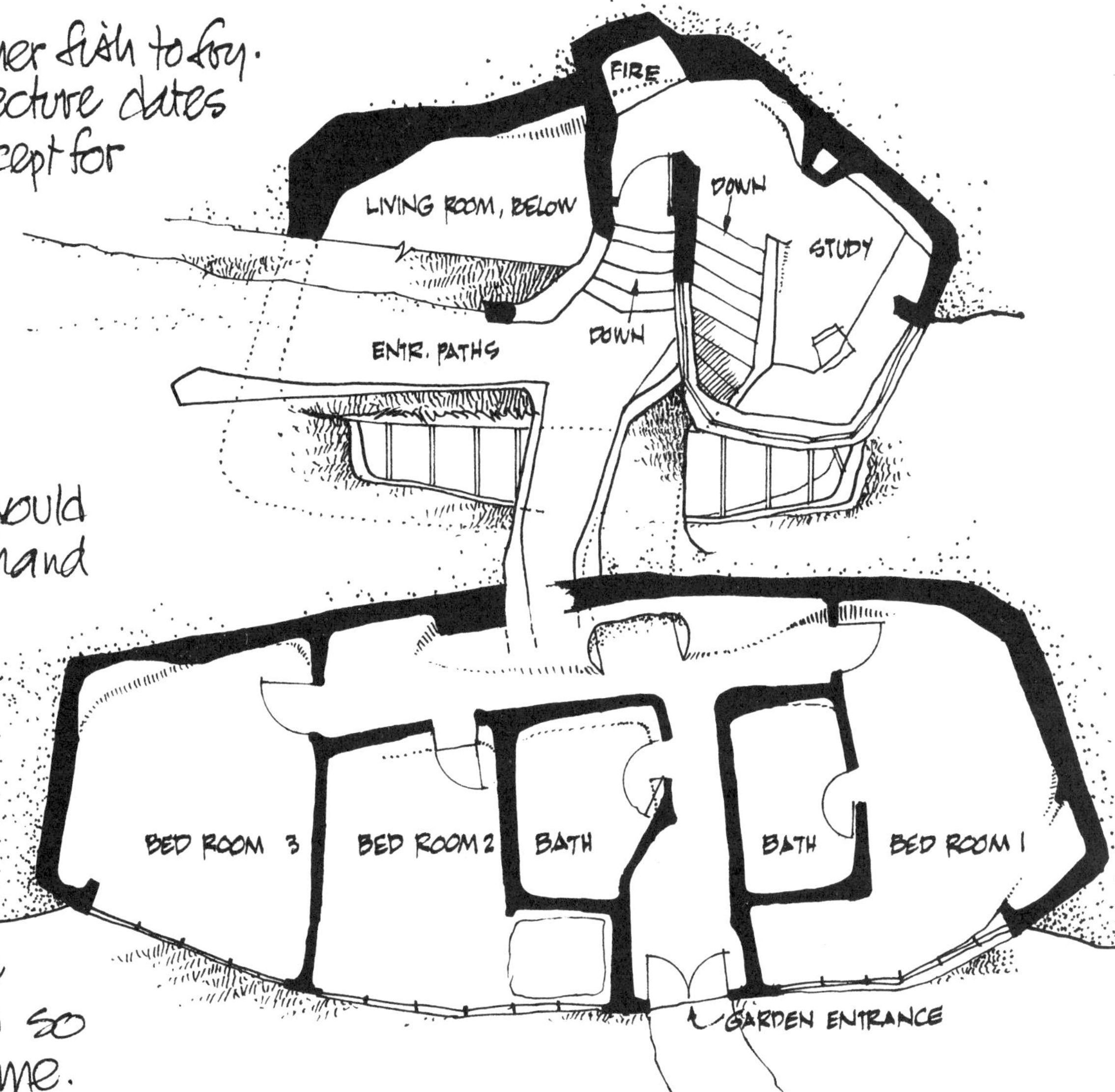

It was, and still is, the most popular of all my underground designs, even though, to my knowledge, no one has ever built it, and, even if he did, it would have to be built entirely by hand and would take years to build.

There's something about the earth-relatedness of these free-form spaces that seems to tug at a lot of us, just as the Couelle designs, from which I borrowed so much of this one, tugged at me.

I use its appeal to convince myself that we have more than an intellectual interest in getting closer to the flowering earth. I think our roots are in it, too.

One result of the P.A. article was a commission for this house in Cherry Hill, New Jersey, my home town from 1949 till 1977, when it had become so bad I had to flee.

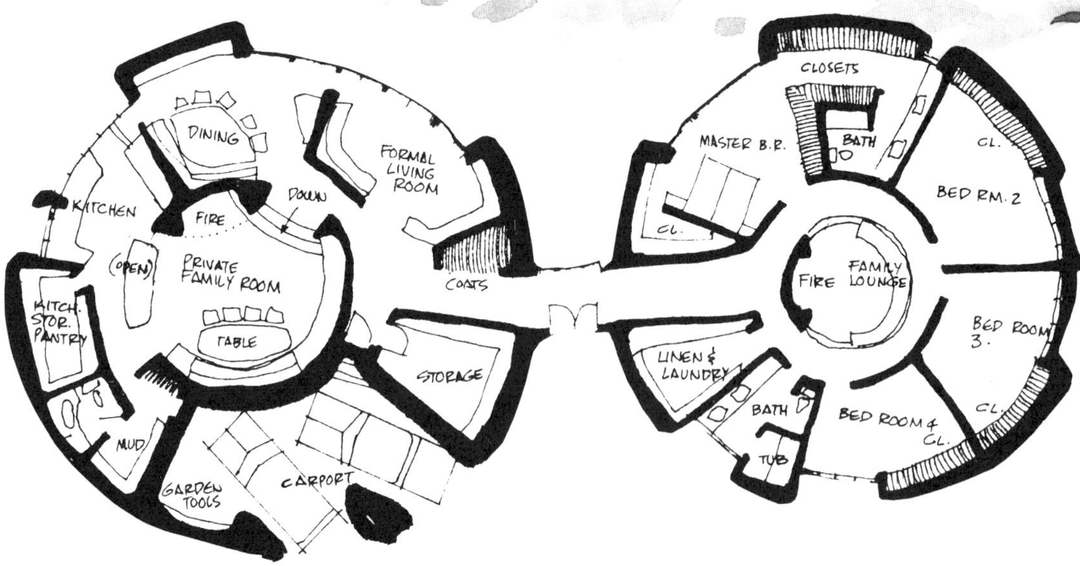

Mr. and Mrs. Lem, the clients, were generous in their reactions to the design but the land purchase didn't go through and the project be-

These were to be classrooms at an environmental center in South Jersey's wet, low cranberry country. "Sorry, no funds," said the state.

And you'll read all about this little chapel when you get to page 19.

→ came an orphan. It was the first of many orphans growing out of that article....

I didn't know, in 1965, that ice pressures in rooftop soil could crack retaining walls, causing all kinds of leaks and structural damage. Fortunately for all concerned, I found out the easy way, by reading about it, so this Mansard-façaded roof-garden building never had a chance to embarrass me. (It was never built.)

I had been sure that my developer friend, Bob Scarborough, would leap at the chance to pile tons of earth on top of his new office building, even though no one of his acquaintance — or mine either, for that matter — had ever built anything like it before.

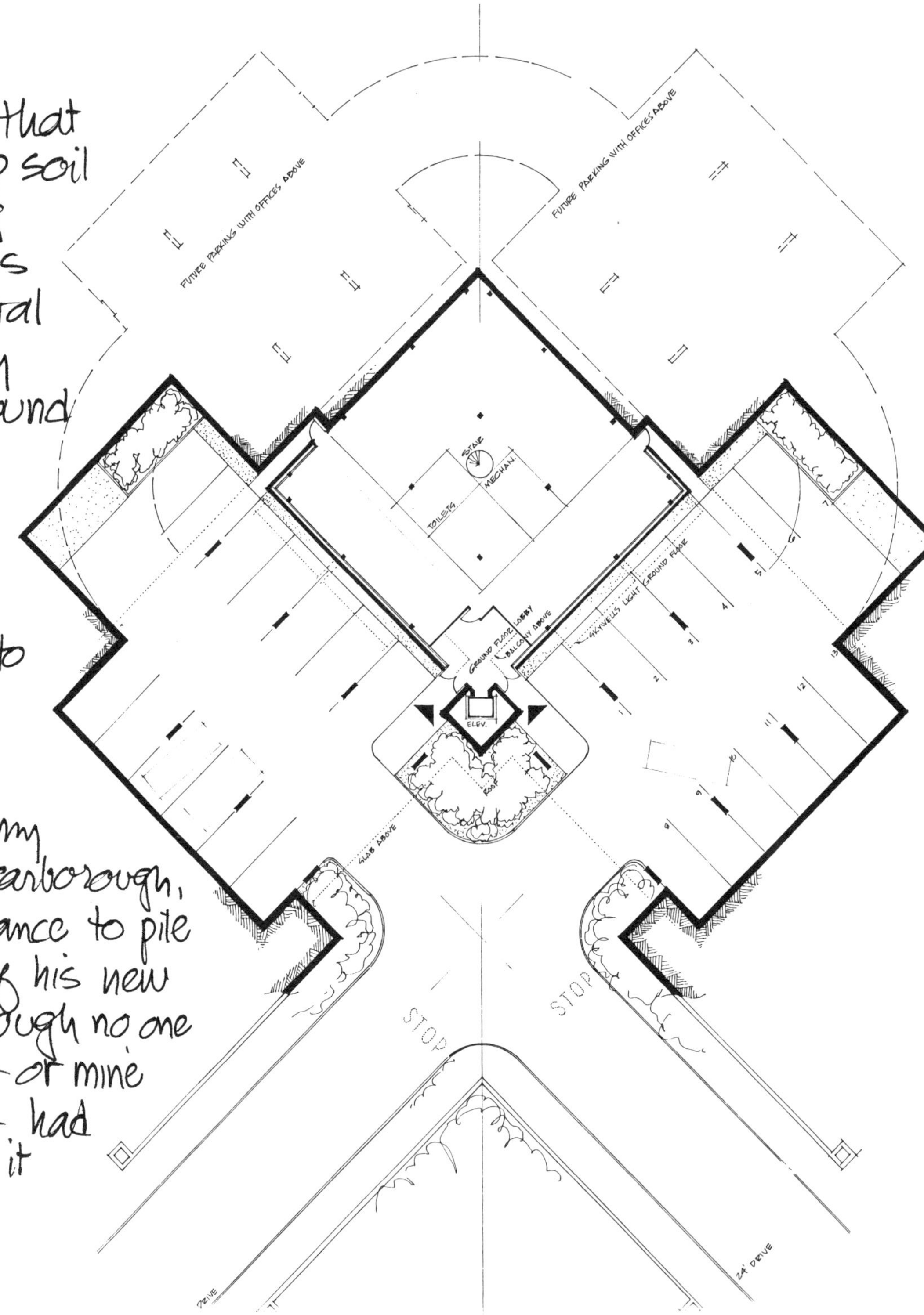

Kappy Wells is my number one child so she gets a whopping 88/93rds of this page just for being so nice, artistic, beautiful (Oh, Dad, cut it out!) and tolerant. I expect her to do greater and greater works of nature sculpture.

Bob was too smart to fall into a trap like that. He paid me and thanked me for my efforts and built a conventional structure that served him and his growing staff well for many years.

As is obvious from a glance at this design, the ideas of heat loss and of solar energy had not yet entered my consciousness.

It was just another exercise in testing the possibilities of building in the earth, and, as such, it probably was of some value to me.

This cross-section

And then there's John, my number one son, whom I see only once a year, but what a heart-warmer, it is to see that bearded bookman whose world is, as it should be, one I cannot enter. To John I dedicate 12/13ths of a page for whatever use he cares to make of it all.

is an invitation to disaster – thermal, structural, and fluvial, not to mention architectural... a wasteful above-ground building dragged, against its will, into the mud of South Jersey.

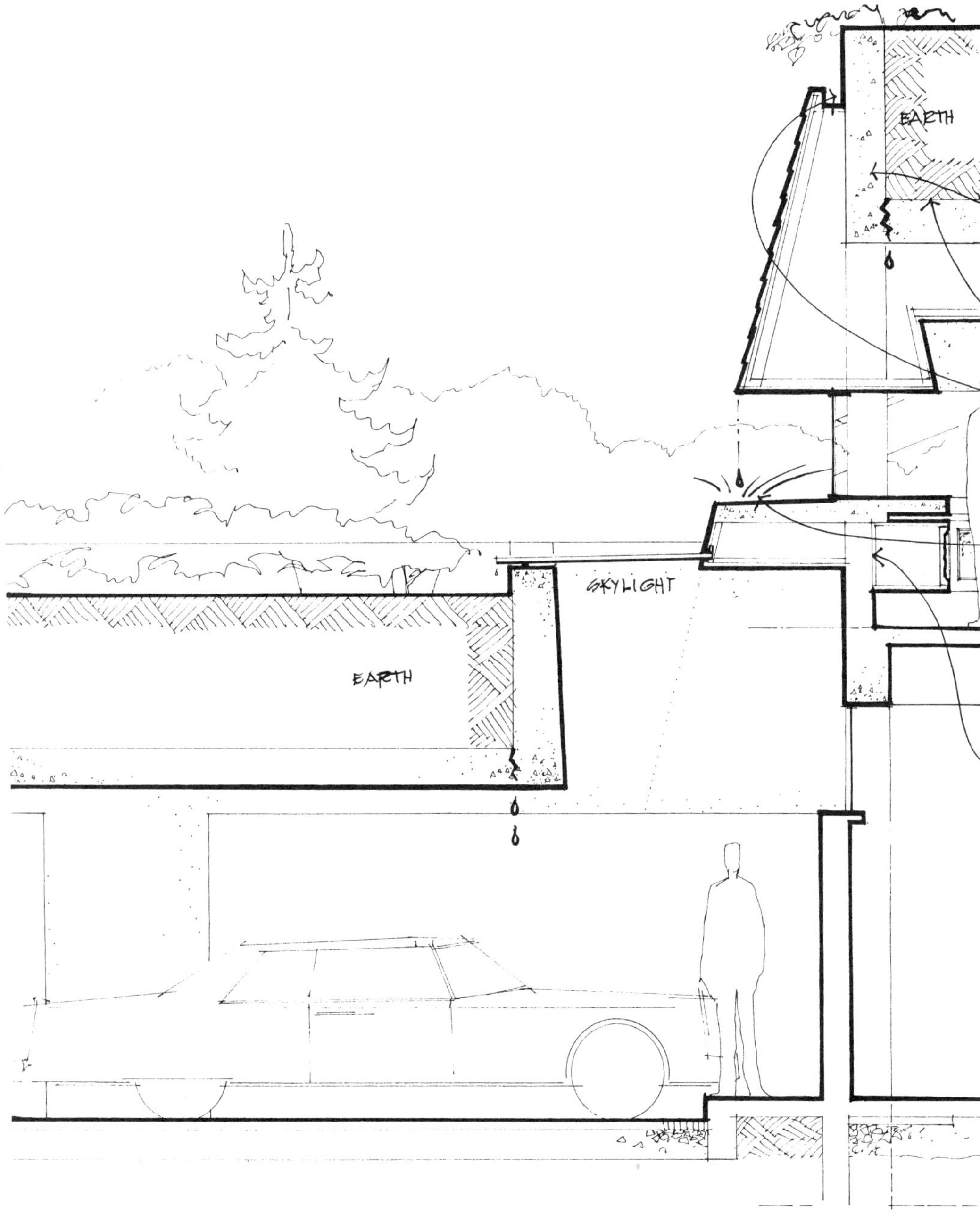

The parapet would have cracked, and leaked as ice in the earth expanded,

heat would have poured out of the uninsulated building,

rooftop rainwater run-off, not to mention mud and plant debris, would have choked the gutters,

splashing rainwater would have kept the windows dirty.

ice-cold floors would have been torture to feet,

and room-air, condensing on cold surfaces, would have formed ice, and puddles, behind every built-in, above every ceiling.

But the whole thing would have <u>looked</u> presentable, possibly even appealing, to passersby unaware of the internal flaws. The then-in-style Mansard roofs, the great planters, and the horizontality of the lines (in contrast to the growing jumble of competing horrors along the highways) might have been judged acceptable, in that way beginning to institutionalize a whole family of design that needed far more study before edging its way into the architectural mainstream.

Lyndon Johnson was president, and roadside beauty was the only issue that even came close to being an environmental one. Gasoline and oil were cheap. And a faraway war in the jungles of Asia was just starting to trouble us. We were still living back in the Fifties.

The mounting problems of waste and pollution exemplified by the American way of life, the destruction of wildlife habitat, the threat of nuclear power, the plight of black people and other minorities, the energy "cost" of materials, the erosion of America's farmsoil, the suffering of starving people, even the idea of harmful food additives... these and a score of other issues so well-known to us today were known to few of us in the mid-60s.
Only after failing for years to see it, did I finally awaken to the damage that buildings do to living land. Of all the other problems I was as

Rounding out the Wells kids there's Sam, the architect, explorer, reader, dreamer, designer, who shares with his brother a sense of humor that can go by so fast I don't even see it go. I don't see Sam too often, either. Sam: 85/90—his—to you.

innocent as a newborn babe.

Thank god for subtle influences! Because of my underground proposals I was starting to correspond with, and to meet, men and women who were able to see the big picture and to transmit a little of it to me.

Solar energy! What a great idea!
Percolation beds!
Composting wastes!
Natural landscapes!

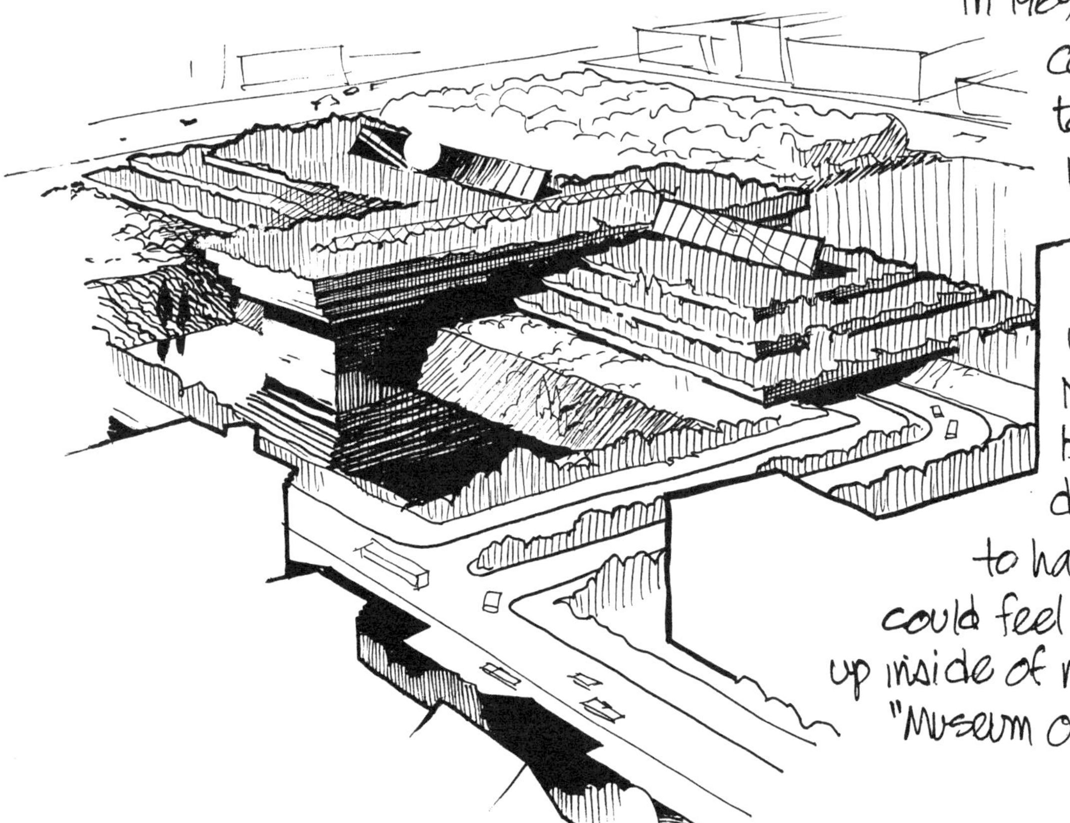

In 1969, the American Association of Museums convened a gathering of great environmentalists like me (hee hee) to deal with the ways in which the museums of America could transmit the environmental mess —and message— to the public. There I was, sitting in a huge conference room in New York, listening to eminent ecologists, biologists, naturalists, and museum directors discuss issues that seemed to have immense relevance for architecture. I could feel a whole new type of design swelling up inside of me, producing as its first fruit this "Museum of the Future" as the centerpiece of our

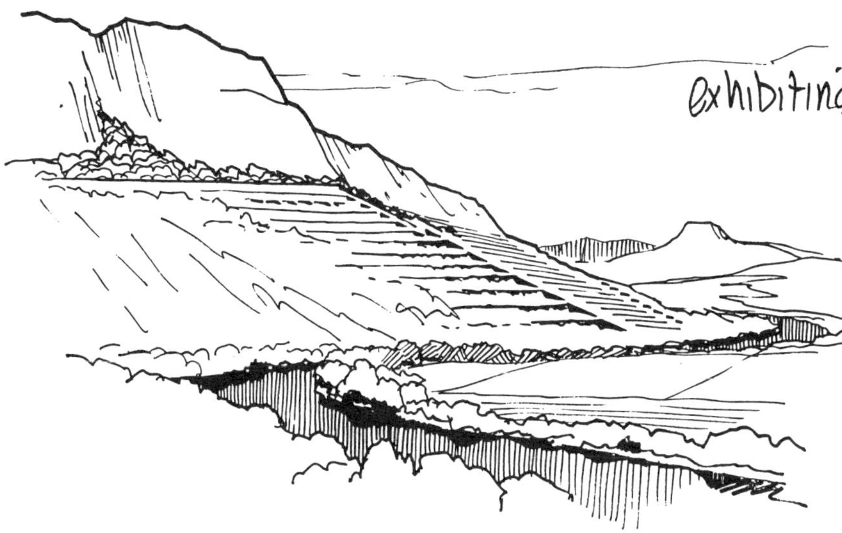

report. The idea was that by exhibiting these things outwardly, museums could say, at a glance, that a new land ethic was here.

The report generated a lot of design commissions. The first was a solar village on the south face of a great mesa in southern Colorado. Almost totally self-sufficient, this complex of solar hillside earth shelters was to have been reached by means of a covered escalator or tramway, up from the cars parked at river level. And then: poof! It all disappeared as the backers of the project turned to other matters and the momentum was lost.

A similar poof! was this invitation to design a headquarters building for the national park at Big Bend, Texas. You can see more details in my 1977 book, Underground Designs (see page 191), but this drawing tells perhaps all that need be told: reverence for, rather than domination of the earth, a sentiment now more strongly expressed by the Earth First! movement.

Meanwhile, back in New Jersey, interest, if not budgets, was growing, spurred by the activities of Earth Day '70.

I wasn't at all sure that solar heating was practical but the people I respected said it was, so I continued to propose it whenever I designed an earth-covered building.

This was to have been the headquarters of yet another New Jersey land developer, a change-of-pace building at which, once the visitor had driven in through the high mound of earth along the highway, he would be in another world, cut off from highway noises and ugliness, in an emerging natural landscape.

It might have gone ahead, but then his whole operation was moved to Washington.

As the seventies began to unfold and energy prices inched upward, things like solar energy and improved insulation began to get more serious consideration.
It was not an unalloyed blessing, for it played to our emerging "me-generation" selfishness. Gentle Architecture was being accepted for less-than-altruistic reasons, and the underlying environmental motive was forgotten, forgotten so completely that it is now just starting to reassert itself, once more for perhaps the wrong reasons.

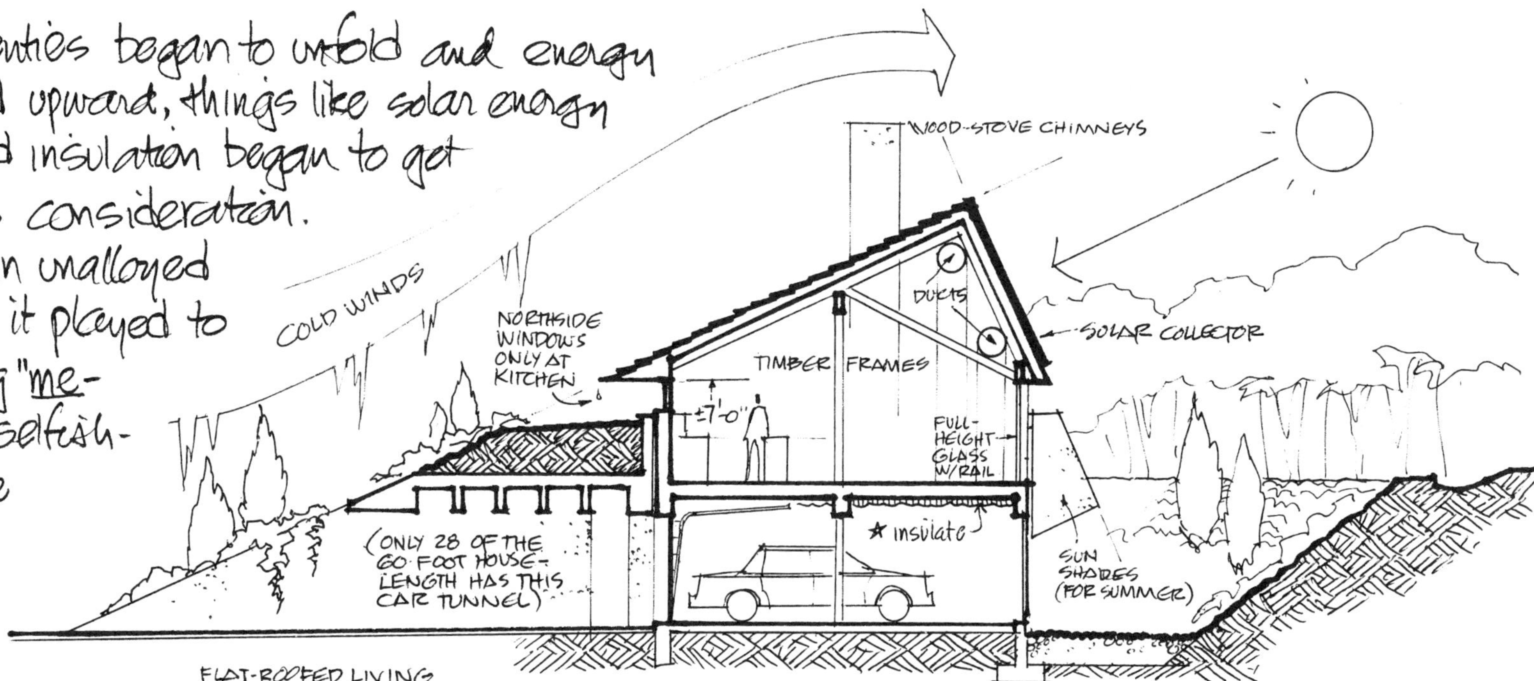

No, this one didn't get built, either. A half-in, half-out compromise design, it was to have overlooked the Chesapeake Canal, but a divorce got in the way and the job was never heard from again.

Jonathan James Kelly, known to his friends as Gene, had a 57-year-old architect dumped on him when he was only 12 but he took it like a man. Now he's a fine man, fully deserving of the 59/64ths part of a page that I hereby dedicate to him.

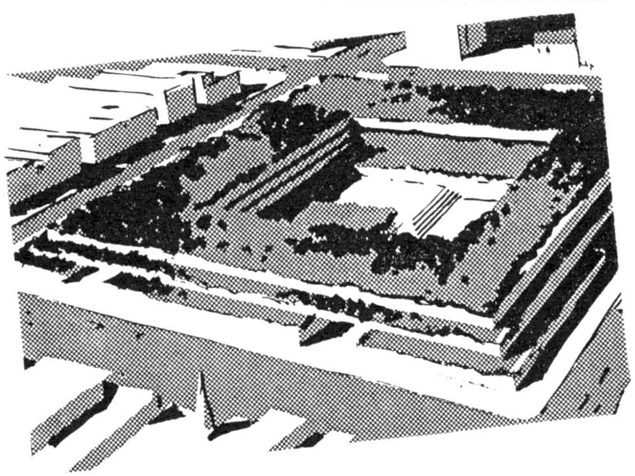

You probably never even heard of The Great Environmental Coloring Book of Life and Death and Architecture, did you?

Self-published in 1971, it was used as my platform, for a while, to present a lot of the ideas that weren't getting used in the real world of envionmental architecture.

No one, apparently, ever used it as a coloring book but it did fall into some good hands, generating a bit of this new architecture... A percolation bed here, some natural landscaping there...

I can see now that these first two sketches came from a) the winning design for Boston's city hall, and b) the heliport at the 1964 New York World's Fair. Number three came from our growing interest in the plight of the poor – mostly black – people – trapped in the worst parts of dying cities. They were – and are – forced to breathe the worst of our air and drink the worst of our water.

Bobbie Wells I liked from the start, and the liking has grown with the years. In honor of that growing affection I hereby dedicate 3/195 plus 20/27ths of a page to her. What a thrill it must be. Bobbie: be sure to read this book. You're in it somewhere, I promise.

Out beyond the suburbs, where the air and the water were still relatively clean, a group of businessmen raising money for a Boy Scout camp chapel actually went for the idea of covering its roof with earth as an expression of reverence for life. The idea was publicized, money was raised, and plans were drawn, but when it was decided that the scouts themselves should build the structure the roof was changed to one of logs, 2x4s, and plywood, and the design was diluted beyond recognition. Whether the scouts themselves had any part in the decision-making process was never revealed to me. I do know that the percolation-bed floor of the chapel was abandoned, asphalt or concrete being the replacement.

Well, there's plenty of percolation out in the woods. It was more for a gesture than for a need. It's in built-up areas, where the run-off from buildings and parking lots can get into 30,000-gallon quantities (for every impervious acre involved during a 1" rainstorm) that water conservation measures are so badly needed.

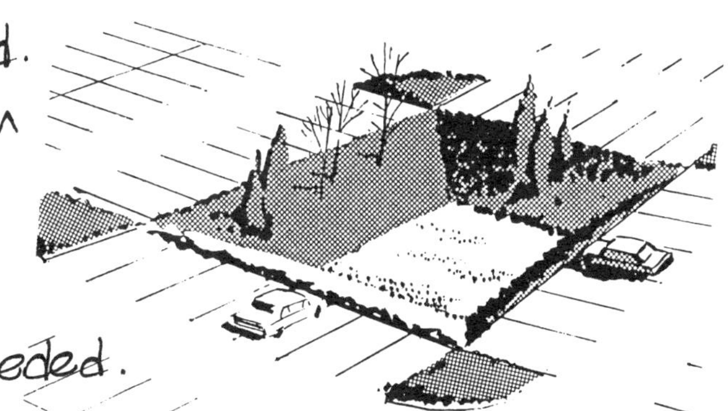

Two more pages of sketches from the coloring book... the subject here: entering the earth; how you go about doing it appropriately. It shouldn't be like walking into the shopping center or the bank. It should be, somehow, an act of humility.

The wild creatures of the earth know how to do this. Their entrances, while created, admittedly, for defense, have an appropriateness we have never matched. When <u>we</u> build entrances designed primarily for defense they look even more forbidding and destructive than our conventional entrances, which are bad enough.

These 3 sketches were done as design proposals for a state park visitors' center in Pennsylvania. Building a naturalistic-looking entrance that wouldn't invite all sorts of leak, dirt, rodent, and security troubles would be a challenge, as would the idea of creating something that didn't look as fake as some "natural" buildings in zoos.

Stewart Rawlings Mott has been my son for only 2 years and yet he's a tad over 50! (How can that be?) This super-smart man of so many interests has a giant chunk of this page, now, in his name, which I believe you already know.

There was talk, in the early 70s, of building a huge jetport in New Jersey's "Pine Barrens", the forest that covers much of the center of the state. There was talk of square <u>miles</u> of rooftop and paving areas, and of what the run-off from so many chemicals would do to the forest's aquifer from which so many New Jerseyans drink. My proposal for a jetport at which everything but the runways was underground cut the paved areas by almost 90%. But then the state stepped in and bought the Pine Barrens as a conservation area and the great underground airport was no more. How long the state will hold out in the face of development pressure pinching it on both sides, from Philadelphia and Atlantic City, is an open question.

Another nature center that never got going, was this. It had a rooftop pond and stood in a pebbled courtyard.

A few months later, realizing it was time to put up or shut up, I built an architectural office for myself using the pebbled courtyard idea.

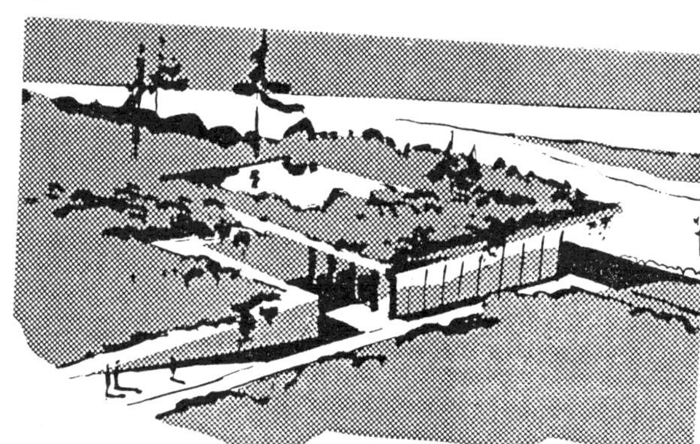

Almost all of this page is dedicated to a young architect named David Marley who may, before too long, formally join this doughty band of pilgrims as it waddles down the corridors of time, spreading goodness and truth along the way.

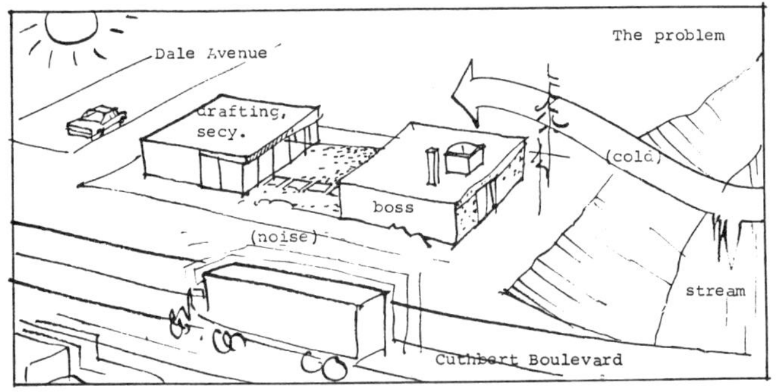

There was a tiny lot (60'x90') behind my old – dare I say it? – above-ground office in Cherry Hill, New Jersey. Used by the state highway department as a storage yard for various paving materials, it had very little going for it. On one side was a 6-lane freeway; on another a badly-polluted stream.

A conventional building – or pair of buildings – out there would have been all too appropriate in that part of New Jersey. Who would have complained about two more ugly boxes along that noisy highway? But I couldn't bear the thought, so down we went.

That's when the surprises began. The first one was that the open courtyard was actually quiet in spite of the truck traffic only 20 feet away. Another was the absolute silence of the innermost rooms. Another: discovering that some of the small animals – turtles, rabbits, moles, and mice – that moved back onto the newly green site couldn't avoid falling from the tops of the 7-foot walls, sometimes to their death.

Jessica, my number one granddaughter: how much of a page shall I dedicate to you? Would you like a bit, a chunk, a hunk, or a lot? It's up to you. I hope you pick the latter because I think you're going to do great things with your life.

Another surprise was the discovery that earth alone is not an adequate insulation material. Cold weather made ice form – indoors – on surfaces nearest the out-of-doors!

And I didn't know enough to take advantage of the sun's warmth. If I'd opened the building to the south (↓) and insulated it properly it would have heated itself most of the time. Even as it was, however, the building used less heating fuel than a comparable, well-insulated above-ground office building of the same size.

The building not only attracted wildlife, it attracted tourists, publishers, reporters, architects, TV crews, relatives, and just plain well-wishers. As oil prices soared, interest in the little building soared with them, but seldom for the right reasons.

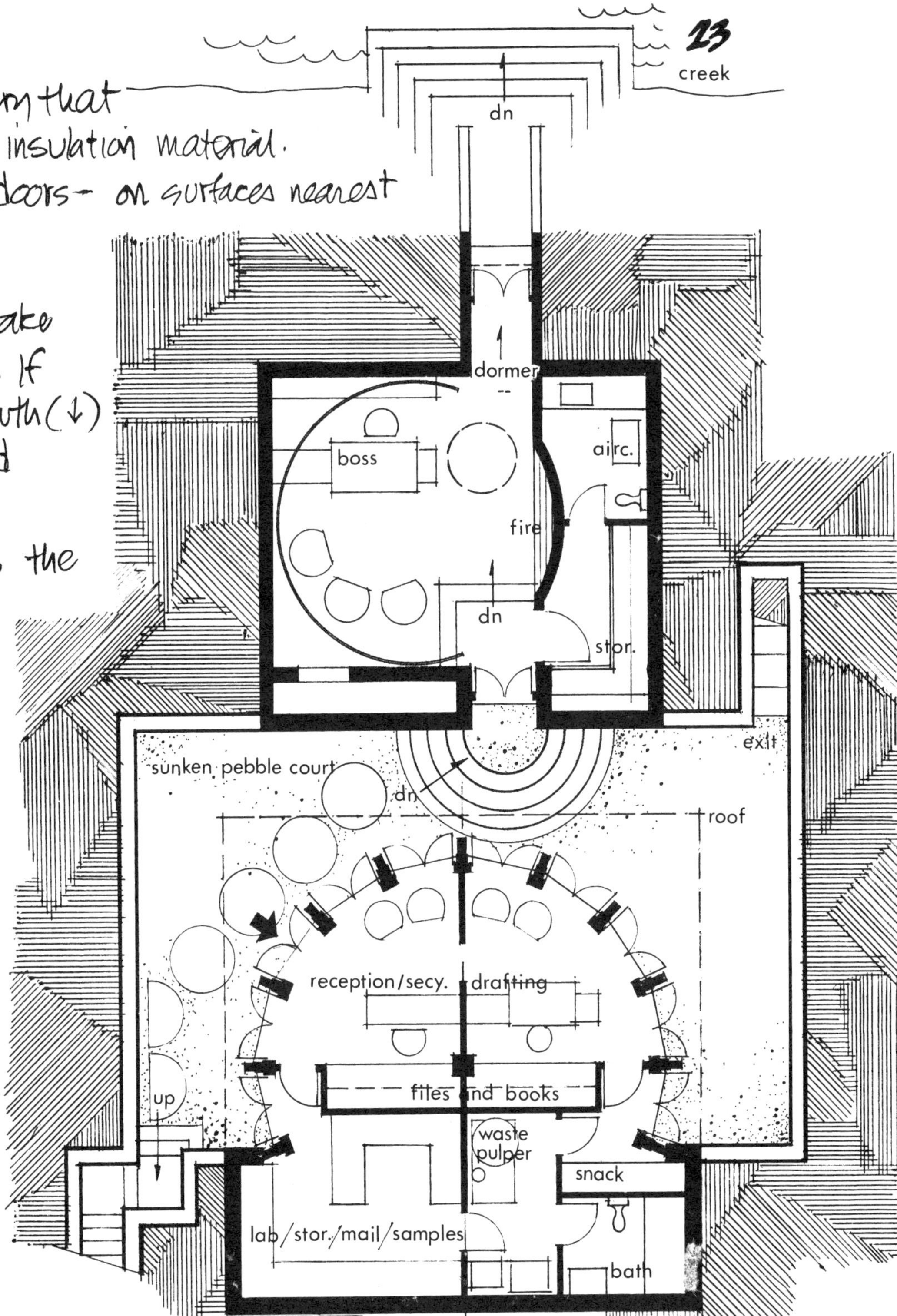

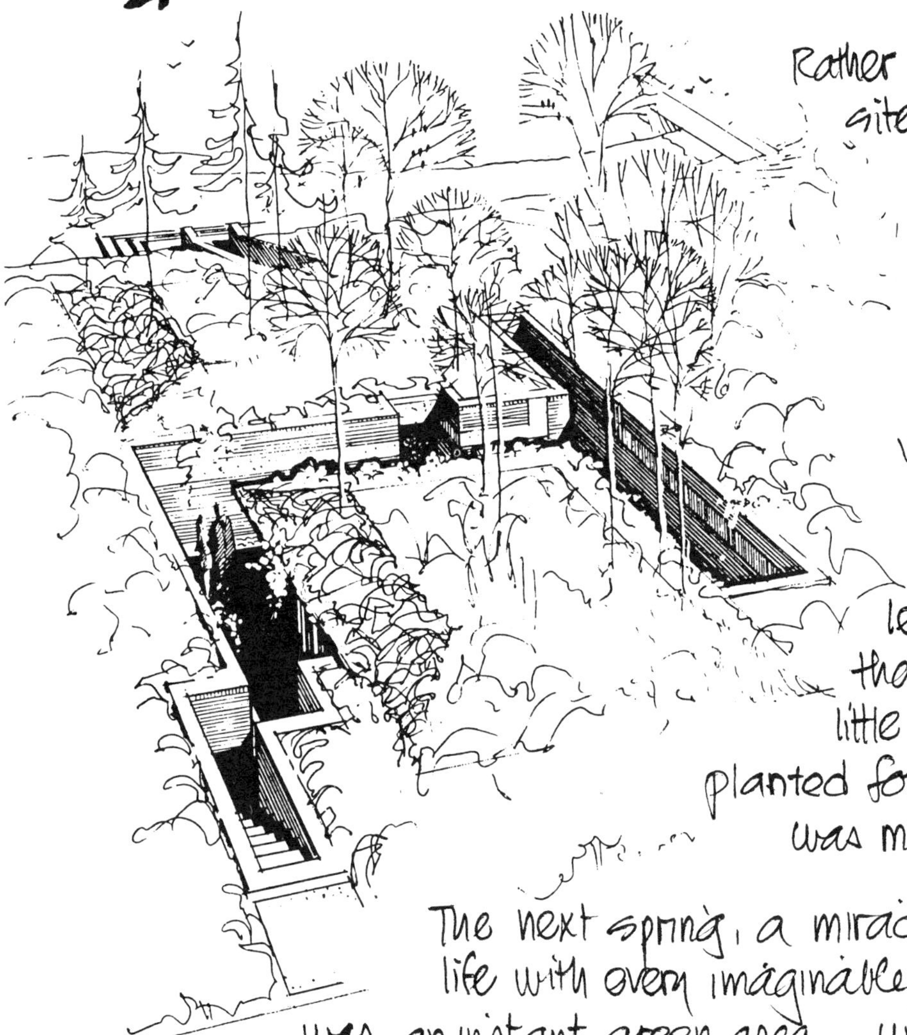

Rather than bring in topsoil from another site, I decided to see if topsoil could, in effect, be created. Once the brick and concrete structure had been waterproofed (with 1/16" butyl sheets) it was backfilled and covered with the lifeless subsoil we'd stockpiled when the site was excavated. Then, from a nearby town dump, we brought tons of leaves collected from the streets of that town, and spread them there. A little ivy and a few junipers were planted for immediate effect but the area was mainly left in leaves, untouched.

The next spring, a miracle occured: the site burst into life with every imaginable kind of weed and wildflower. It was an instant green area – without poisons, fertilizers, or topsoil. I was amazed at the variety of seeds that had stowed away on board those suburban leaves.

Within just a few years I had a jungle on my hands. If ever there was a happy site in New Jersey, this one seemed to be it. At times the bird-

I have exactly the same kind of expectations for Kate. She's following along, a couple of years behind Jessica, and already I can see her special qualities. May I dedicate everything but the toxic materials in this page to you, Kate Wells?

song seemed louder than the truck traffic. Maybe it was the site's way of expressing gratitude for having been spared the fate of every other site in the area. Others, if they weren't asphalted to death, were kept in a perpetual stupor with poisons, and with mowing.

Now almost 20 years have passed since the little office was built. Now that I live in Massachusetts I don't get back there very often. Who knows? Maybe there's a fast-food joint on the site today. Two or three years ago, when I was last there, I couldn't see a sign of the building from its driveway. There was just the old walkway, disappearing into a young forest.

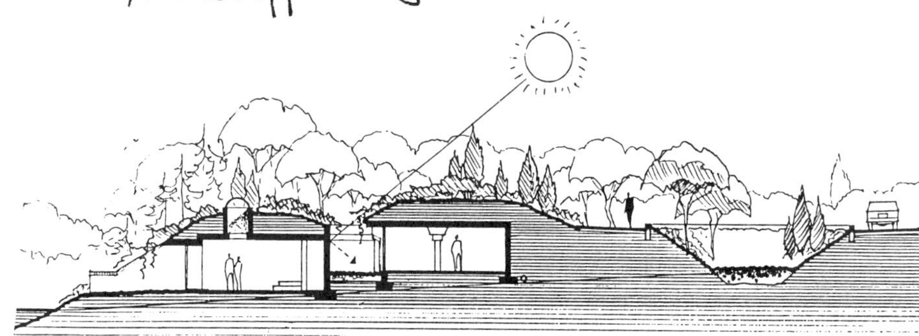

Section facing east

The percolation bed at right is in the center of the driveway loop.

section, facing west

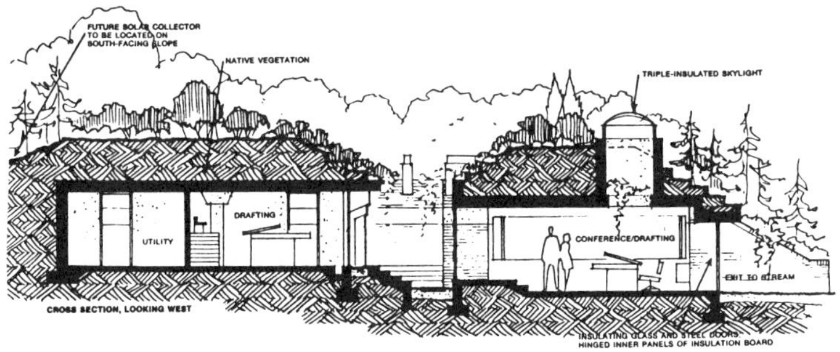

somewhere along the way, I added that "future solar collector" note (above) in an attempt to cover my failure to add the real thing.

Come on in and join us, Samuel Apple Axle Mott. You can't escape the inevitable. A hefty chunk of this very page is set aside in your name. Who knows? Maybe you will someday carry on the work of the architectural underground and do great grottoes.

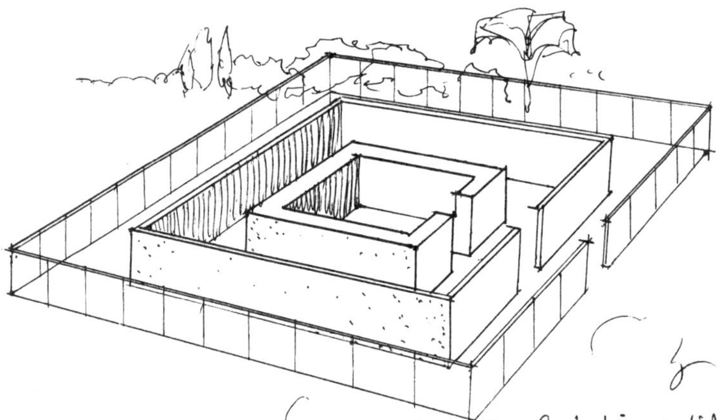

One early series of studies I did involved a many-layered house in which rooms would be peeled off, or put on, by moving in or out, depending on the degree of privacy, insulation, etc. required.

Energy conservation! It was such a hot subject in the early seventies we were all looking for ways to use less fuel. Automobiles were getting smaller, everyone was weatherstripping, and solar heating was no longer a fringe idea for use only by fanatics. Even the politicians were getting involved, offering tax credits for various conservation measures.

A dream of mine was – and still is – to build a box about the size of a large coffin, arrange it in such a way that insulation of various types and thicknesses can be applied to, and removed from, its 6 surfaces, and then crawl in and close the door (on a bitter cold winter night) and see how much insulation was needed to keep me warm, while lying in there, naked, using nothing but my body heat as the energy source. I'm still going to do it. Someday.

To all my cousins, nieces, and nephews: this page* is dedicated to you in-der a group plan that entitles you to a 50% discount at any book-store that's willing to go along with such a nutty idea. (* most of it.)

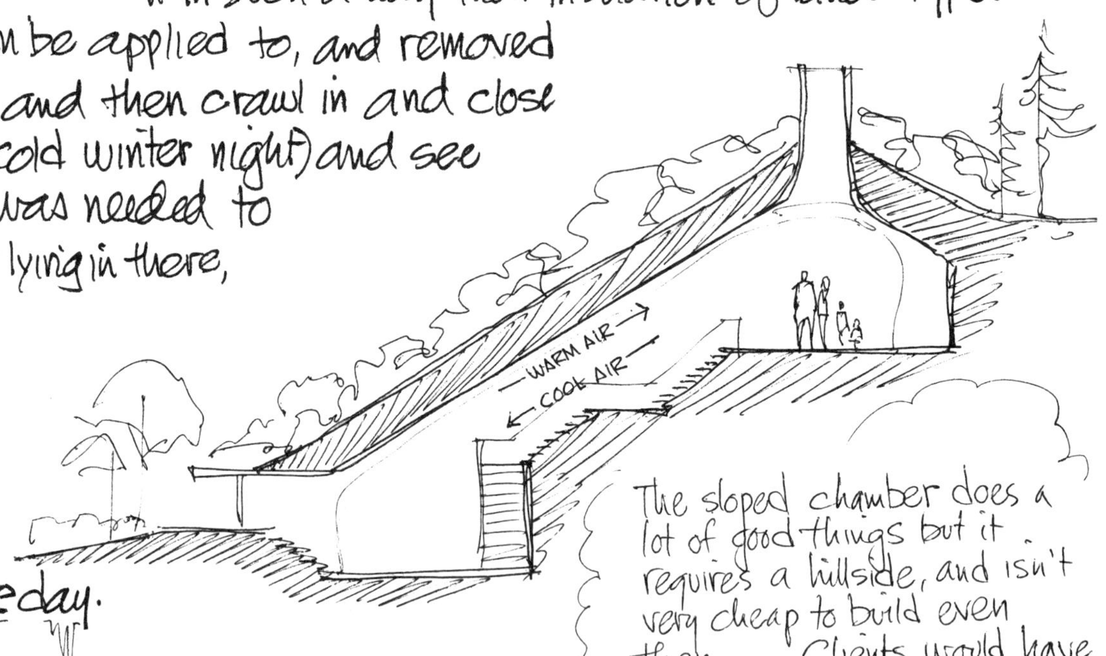

The sloped chamber does a lot of good things but it requires a hillside, and isn't very cheap to build even then. Clients would have to be sort of mole-like, too.

The sketches on these 2 pages were part of a letter I wrote as an accompaniment to the design on the next two pages. It was an idea for an energy-conserving earth-covered house to be built for a client living in North Carolina. As you'll see, it would have cost him a lot of money. He moved to another part of the country before he had a chance to find out.

It seems likely that at some time in the future, perhaps within a generation, we will learn how to heat our bodies (and other frost-vulnerable creatures and things in our buildings) without heating the buildings as well. It would save immense amounts of fuel. The practice of lighting only the surface at hand (instead of lighting an entire room) is called task lighting. Perhaps the practice of heating only bodies and other things requiring heating should be called task heating.

Heated clothing has been proposed, as has heated furniture, but it is obvious that the ideal would be the direct heating of the body. We already have such systems built into our bodies but as we become less and less active we require some sort of supplementary system. There are few highs equal to that of being made warm, in cold weather, through vigorous exercise. Maybe task heating will offer similar rewards.

Frederick R. Johnson of Galesburg, Cos Cob, and South Orleans is my only uncle, and what a nice guy he is! Seventeen years older than I, he looks like my brother—or my son—so to him the top 9/10ths of this page is dedicated.

surely I didn't mean to propose that there be hundreds of square feet of lifeless pebbles around the house. Maybe the jiggling lines were meant to suggest mulch, or some sort of ground cover.

In any case, you've got to admit that winter nights could be mighty snug and quiet up there in those rooms above the main floor, with nothing but the sound of the wind generator to break the silence.

I never did a floor plan, apparently. These drawings were sent off to the client, getting an enthusiastic reply, but then, as I said, the job was left to die... (just as the unwary visitor might have done had she come a-running in along the entrance walk, at left, smack into the glass wall!)

LOOKS LIKE HOME ON THE RANGE OR SOMETHING, DOESN'T IT?

A wintertime cross section and a summertime cross section...

This was in 1972, long before I'd heard about the value of pulling heated air down from the top of a building (p.81) before it can escape through the roof.

Might have been a tad cold and damp, down there on that lowest floor during the winter, even in the sunny paradise of the Carolinas.

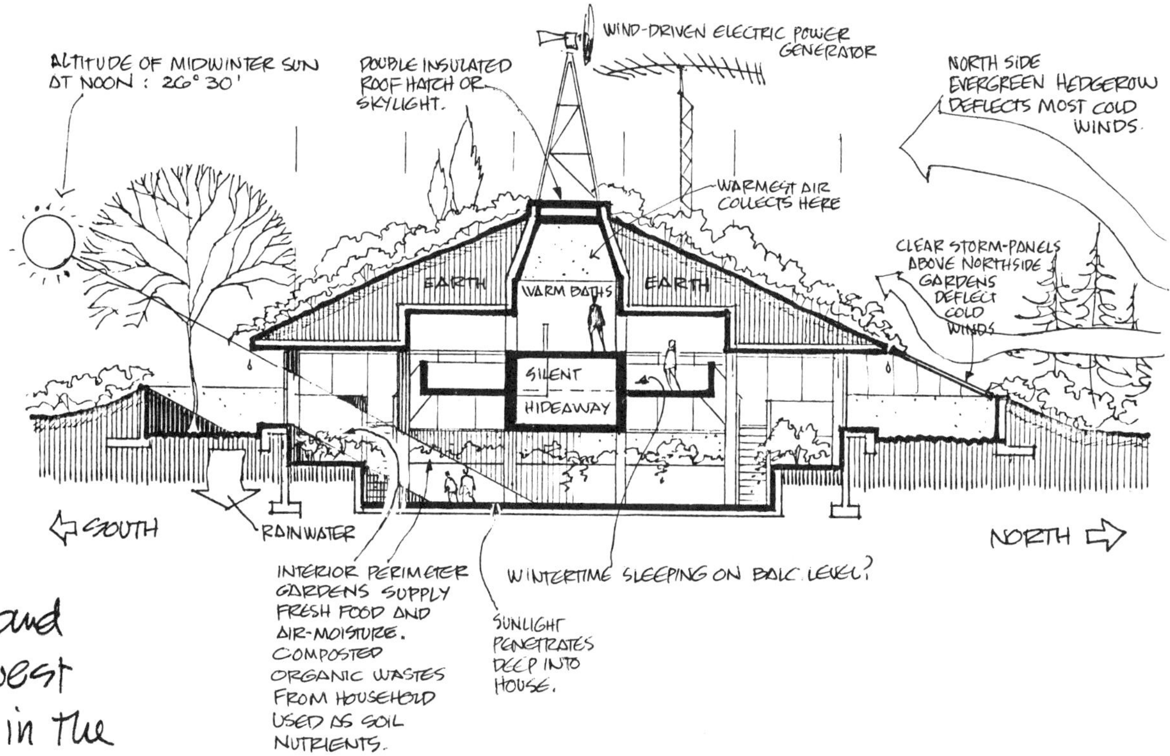

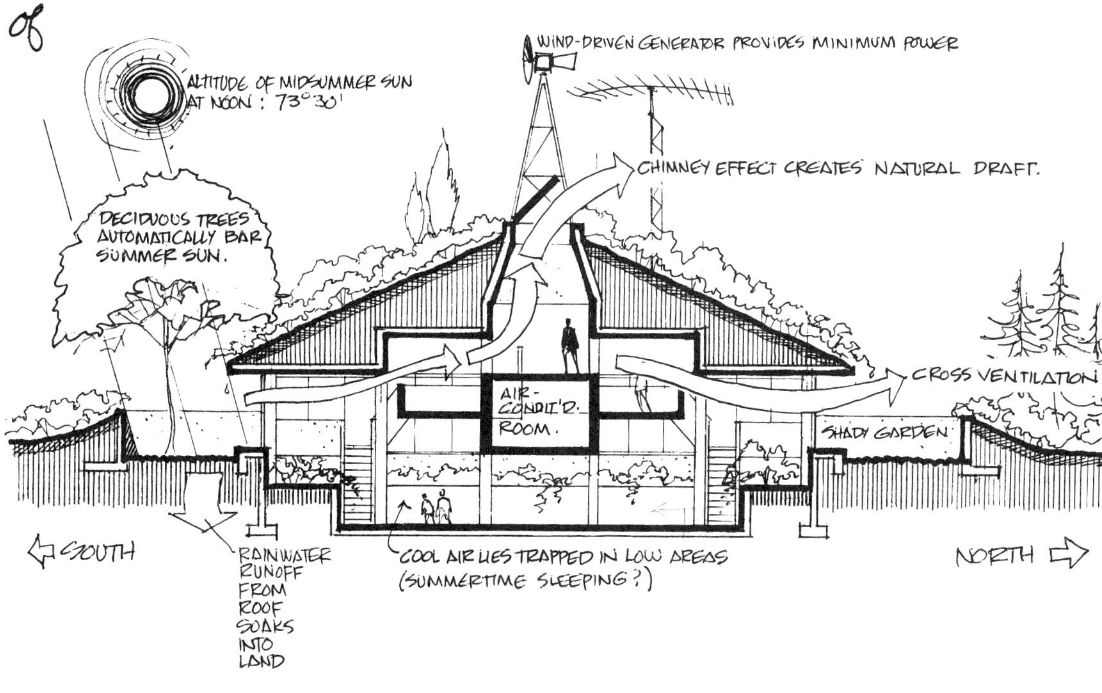

Outside the family now, the dedications go to heroes and friends. In the hero department my first nominee is Ralph Nader, and to him I dedicate everything on this page except for two commas and part of the page no.

One of the hot-growth areas in the New Jersey suburbs of Philadelphia — Evesham Township, I think it was — was said to be considering building an addition to its municipal building. The year: 1974. "May I submit a proposal?," I asked. "Of course," came the response, so I dreamed this dream, watercolored it until it resembled an earthly paradise, and sent it in, never to be heard from again. I never saw what they built instead. I couldn't bear to. But I've seen that township. From the quaint farmlands of the sixties it has become wall-to-wall houses and shopping centers, a town without a center, without any kind of relief from the dizzying mazes of streets that seem to go nowhere.

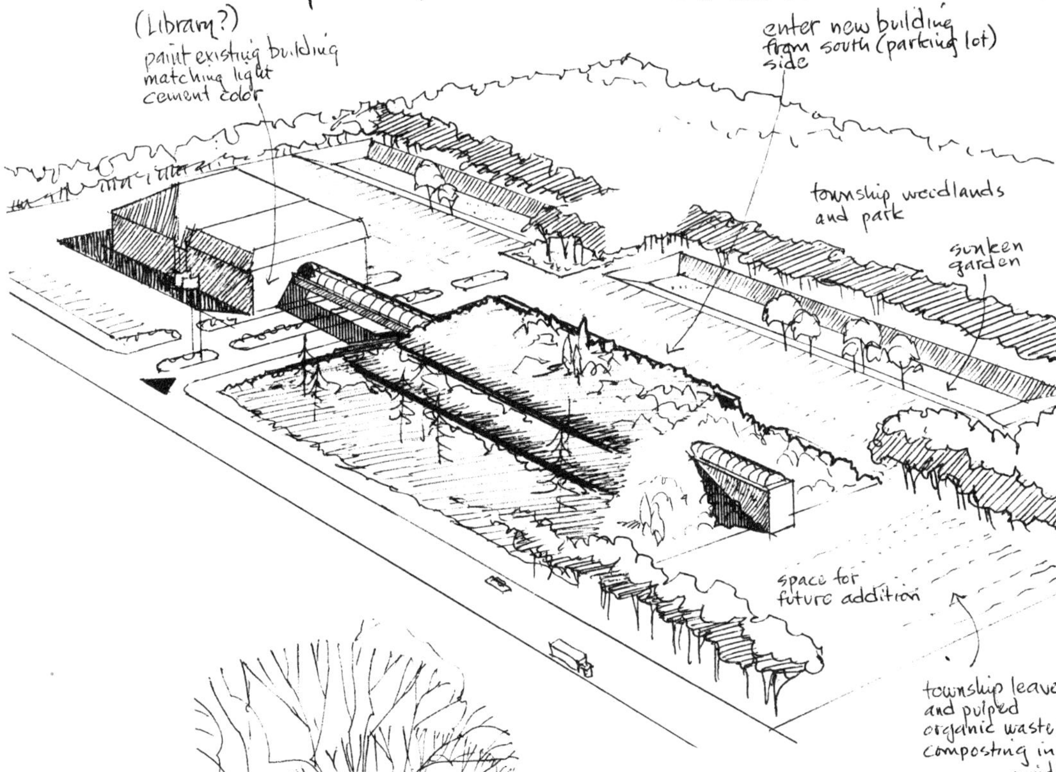

Sour grapes.

Very sour.

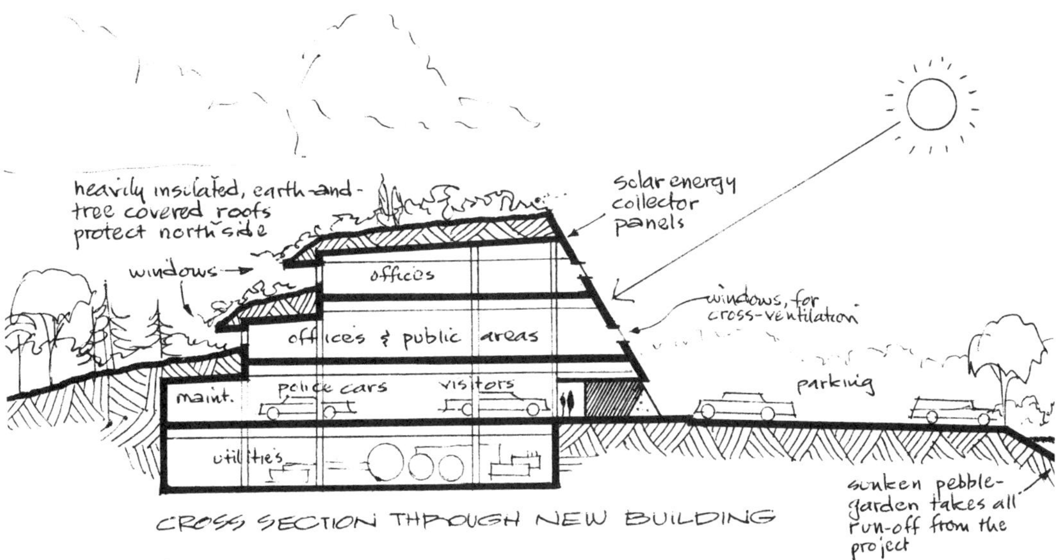

CROSS SECTION THROUGH NEW BUILDING

Unshaded south-facing walls of glass can be murderously hot in the summertime. Sun-louvers or shade trees are needed to shade the glass from the direct rays of the August sun.

If the municipal building were to be built today it would more than likely have large areas of sun-shaded glass on the south side, picking up direct (passive) solar energy rather than have the huge collector panels that were so hot for a decade.

When solar electric panels get to be cheap enough, the buildings may look like these ones again.

DESIGN FOR A COUNTY CHILDREN'S HOME

Craig Waters, I think his name was; what a nice young man. He thought of having a legitimate theatre in the land of the philistines: the endless, look-alike New Jersey suburbs of Philadelphia.

His idea was to build at the local shopping center, making the paving do double duty in this way. He was able to get pledges for quite a lot of money but not enough to justify starting a project of this size, so the momentum was lost and – well, you know the story by now.

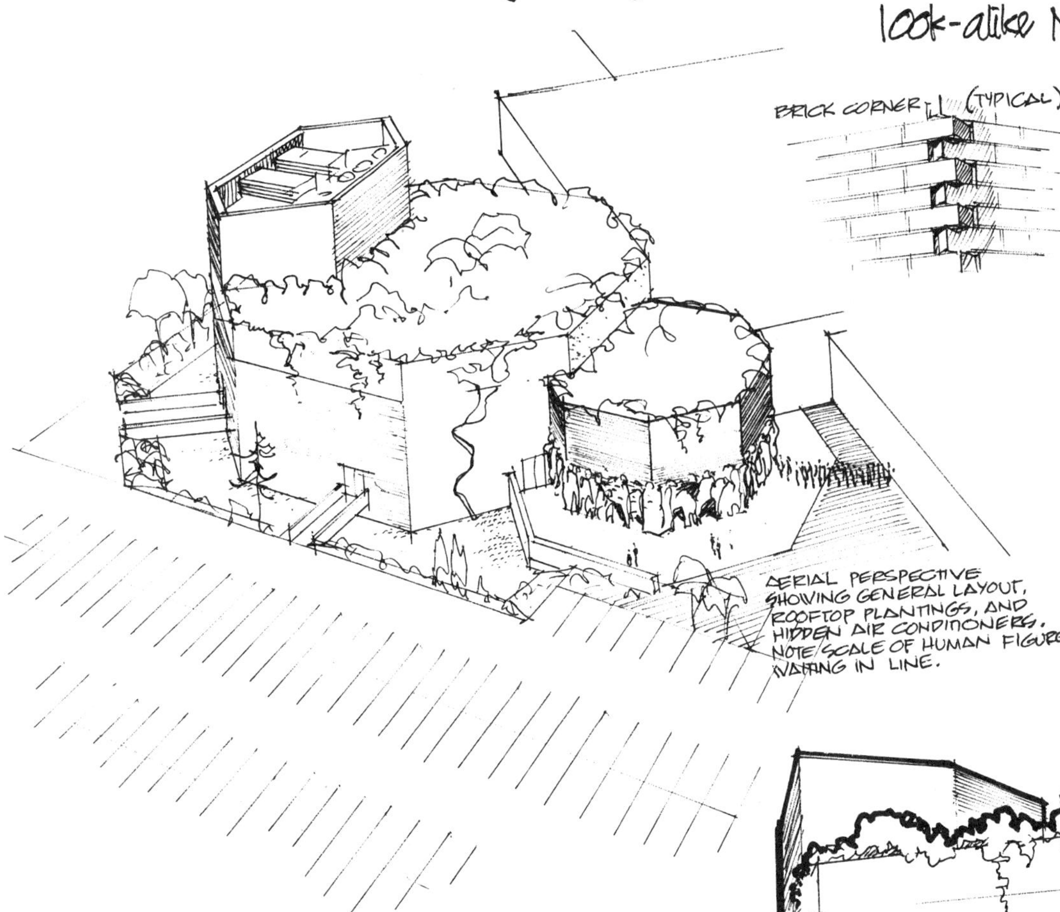

AERIAL PERSPECTIVE SHOWING GENERAL LAYOUT, ROOFTOP PLANTINGS, AND HIDDEN AIR CONDITIONERS. NOTE SCALE OF HUMAN FIGURES WAITING IN LINE.

The fact that my daughter, Kappy, is a sculptor may have had something to do with the idea of using

BELOW: PERSPECTIVE FROM PARKING LOT SHOWING 10' HIGH SCULPTURED FIGURES AROUND ENTRY PORTICO.

Then we have J. Baldwin, my longtime friend, pen-pal, Utah guide, and all-around genius and editor. This dedication plaque will make known to all who pass that to J. I dedicate 13+31/49ths of this fascinating page.

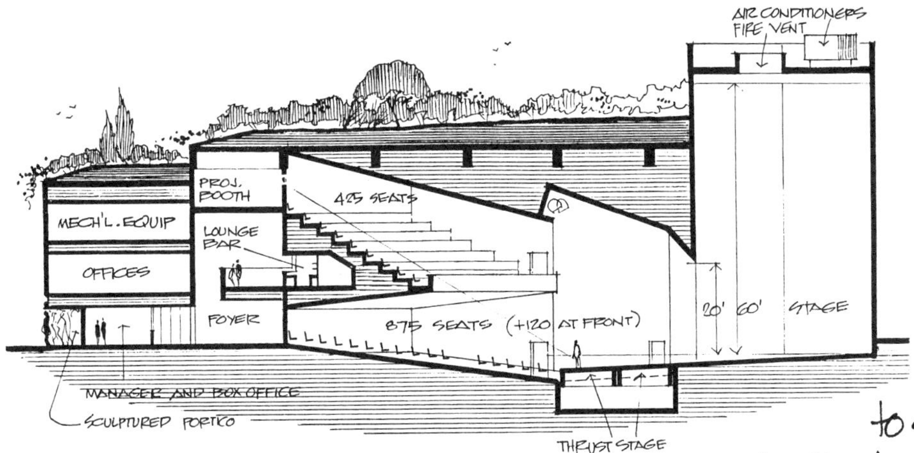

giant figures from the world of drama to support the wall.

This is, of course, stretching the idea of earth-cover a bit far... all that money to support a rooftop garden that only the birds would see – and use.

On the other hand, however, the cost of the added structure needed to carry those earth loads, when compared with the over-all cost of the project, would be almost negligible.

Ah, it would have been a dramatic building, a garden spot in a dreary shopping center.

At least it provided some happy imaginings... and more badly-needed drawing practice.

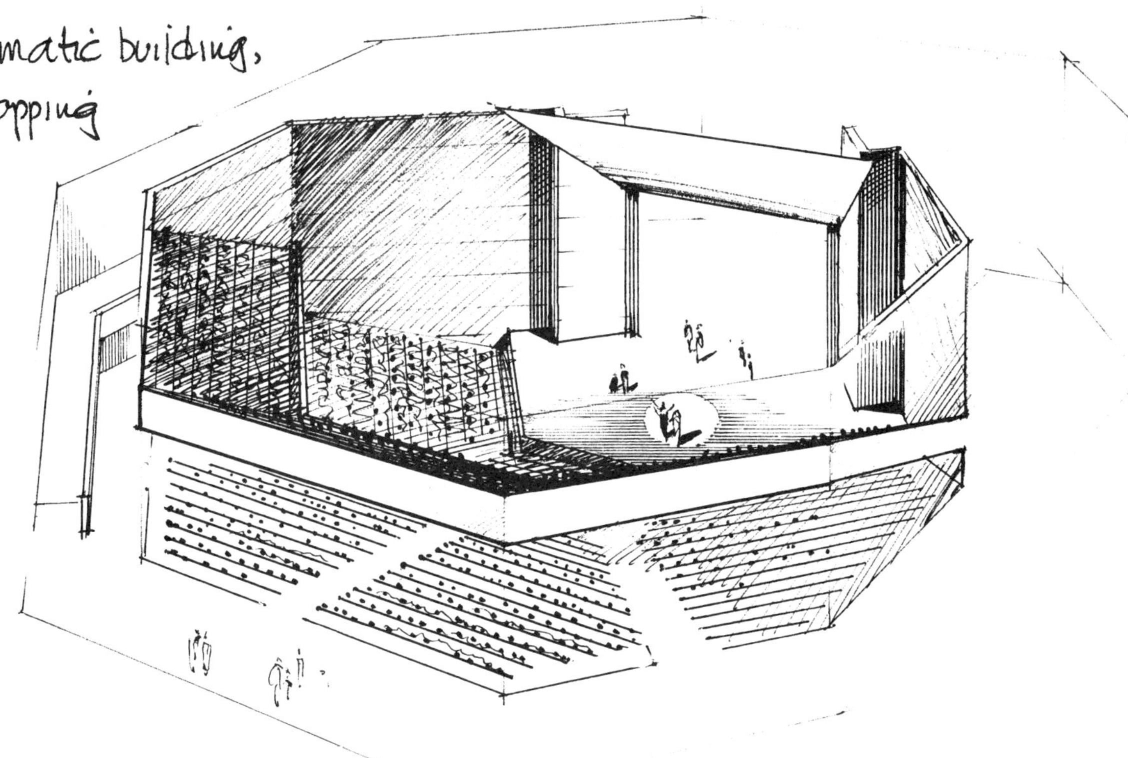

Throw away all the other pages if you must but save these two. I want to be able to say I told you so, along about the year 2090. Here's why: we, and all our buildings and other civilized trappings,

get uprooted, moved, re-rooted, and then uprooted again, ad infinitum, or so it seems. Mobility and restlessness are our middle names. But every time we change our minds the natural world takes it on the chin. You can't clear a forest one day and replace it the next. The same is true of entire ecosystems and habitats. The cycles of change in Nature are too slow to keep pace with our restlessness.

One response would be to take away our wheels, nail us down, make us get back into the rhythms of the earth, but it's not in our nature to live that way, not any more. We're on a quest (some call it a rampage), for what, no one is sure. It seems to be leading us far out into space, far down into the heart of matter, far out to the limits of things. That's the human condition.

Dave Deppen is the Bay Area architect who, when we were in New Jersey showed me how to design a book for self-publication. Now I must wing it without him. To you, Dave: everything on this page but the binding at the edge.

So rather than resist what is obviously the destiny of our race, why not accommodate it, and the natural world as well, by creating totally

changeable environments within permanently immovable earth structures?

Giant earth shelves, built to last for millennia, would allow the wounds caused by their construction time to heal, and to let the natural succession of plants occur. Year after year, the soil would grow richer, the wildlife more varied and abundant, the land ever more beautiful. And, year after year, the restless ones would be free to move, to change their habits, to recycle the old units, and to live without the guilt that now spoils our enjoyment of the world around us, knowing as we do, that all the best parts are soon to disappear.

We could, of course, destroy the entire planet and move on, to another one. We could also destroy the planet and stay here in the ashes. Or we could stand back and give the poor old girl another chance.

Dave Deppen, who edited *Underground Designs*, p.191, used this drawing → on its title page. Hello, Dave.

Crystals of various geometries grow uninvited in the heads of architects. One month, I find myself doing rectangles, the next: circles, followed by hexagons or free-forms. What we see here are products of mixed crystals, no doubt: squares masquerading as triangles. No matter how they actually work, thought-crystals are unending delights to their hosts. I never know, when a design starts to appear, where it comes from, or where it's going, just as I have no idea what words will appear to fill the rest of this page, still blank from here on down.

Can you imagine the kinds of fencing that would have been required to keep all those crowds of New York kids from climbing up over this little exhibit? Jim Oliver, the Aquarium director, was terribly concerned about the destruction of Jamaica Bay, hemmed in as it was, and is, by JFK airport and every other kind of urban development. He wanted to have something, even a little exhibit, to show aquarium visitors the kinds of treasures being crushed out of existence by all the mindless growth and its attendant noise and pollution.

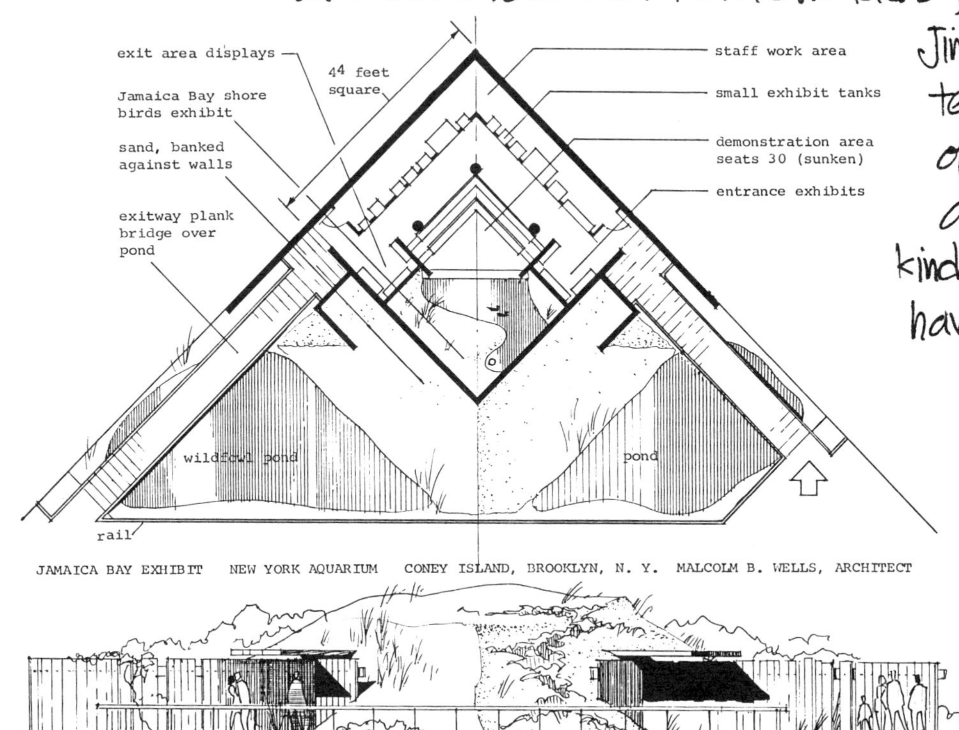

JAMAICA BAY EXHIBIT NEW YORK AQUARIUM CONEY ISLAND, BROOKLYN, N. Y. MALCOLM B. WELLS, ARCHITECT

But the funds never materialized.

A health-clinic director and her airline-pilot husband once took me to a magically beautiful site overlooking miles of Amish farmland. They wanted to build a solar earth shelter in the hillside at the edge of the farms.

I can't remember, now, why nothing came of it... a big little one-bedroom house.

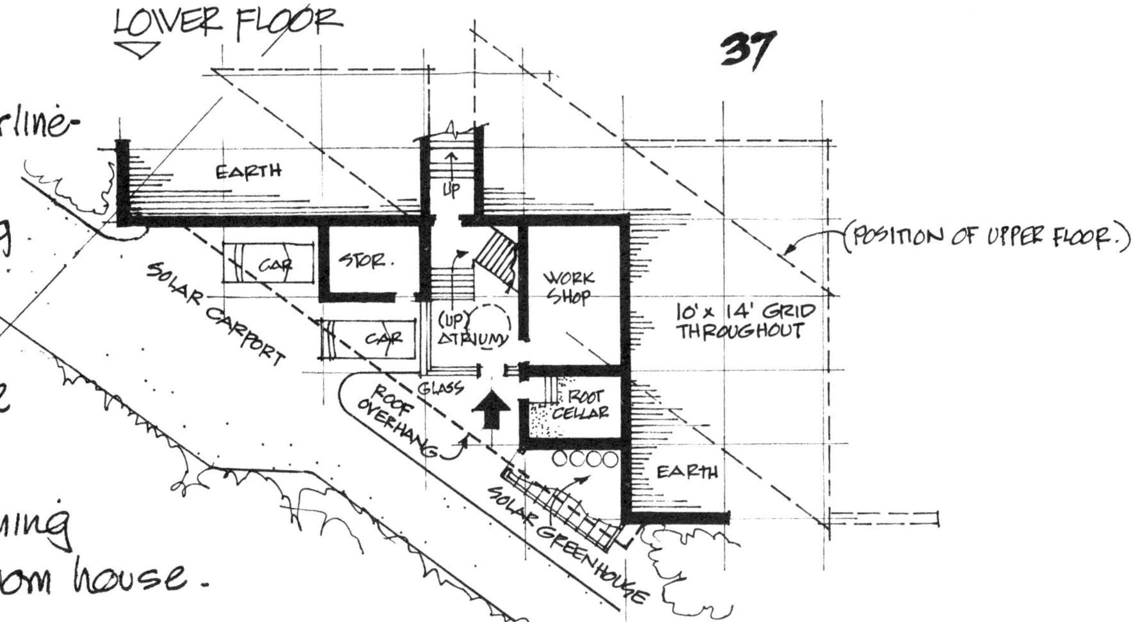

I've presented these plans, as you'll see, in an unconventional way: lower floor above the upper floor. I'm not trying to confuse you, I simply wondered if, in order to better understand whats going on, you should come upon the lower floor first.

(I suspect that this is a failed experiment in graphics.)

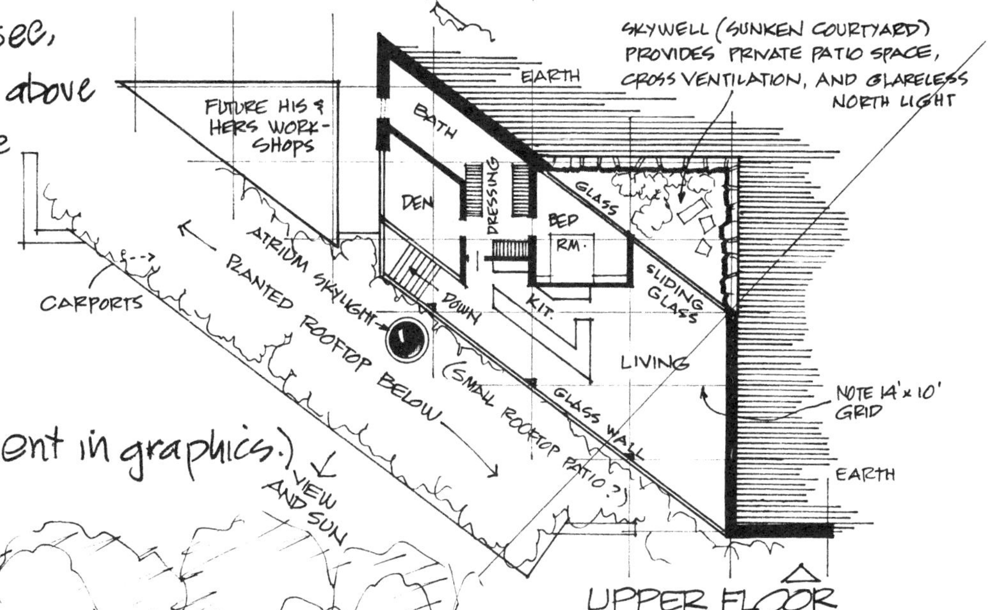

Now. what is all this about? These sketches turned up somewhere in my files and I decided to include them. But why? This first one relates, somehow, to the stepped-up rooms on page 26. The highest room would certainly be the warmest but unless its excess heat could be pulled down and stored in and around the lowest room, the warmth would find its way out through the upper-floor walls and windows. Has anyone built a series of rooms like this as a thermal experiment*?

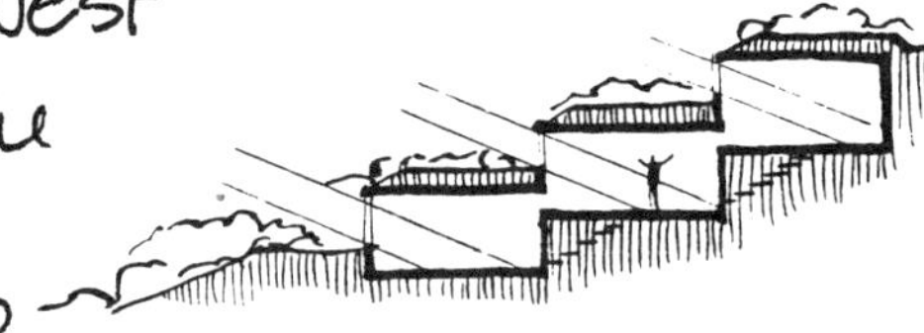

This next one has an upper room, too, but I doubt if it was done for thermal reasons. Looks like a good spot for a secret pagan ritual... sunlight illuminating the altar at certain seasons... It doesn't matter; the point is that the building could be a beauty.

There simply isn't anything like earth cover to endow a building with a sense of eternal appropriateness. Take a look out here between these massive columns. Can you see, out there in the distance, the bands of windows that identify the far-away hills as human habitations? ...terrace after green terrace, down the slopes, housing thousands of people without displacing the plants and animals entitled to share the land with their human kin.

Paul Giambarba came to see me on business 10 years ago and we've been friends ever since. But he and Fran have moved to California! What kind of friends are they? Good ones, I say, and to them I dedicate a huge portion of this very page.

* "Yes," said my friends at Raven Rocks, "remember the Karen Terry House?" (The Passive Solar Energy Book. Mazria. pp 42-43.)

Not many little furry creatures live here. A few mice, perhaps, and Velda in her new mink coat. The average underground house in America is an architectural horror – the worst of suburban consumerist values under the worst of suburban landscapes: the toxic green* lawn. Energy is conserved here, all right, but strictly for the wrong reasons. They use the savings to buy more gadgets. First it was the CB radio, then the video games, the computer, the VCR, and the compact disc. Now it's the ever-helpful fax, linking them to all the other people who share their tastes and values.

Here's a more altruistic family. They saved their money in the hope of saving the land but when that pile of dirt hits the roof... crash! The land will land in the living room. You have to be careful when you place earth on a roof, even a carefully-designed one (and that's the only kind to have).

Construction loads are far more severe, sometimes, than the long-term ones.

*Thanks to Milwaukee naturalist Lorrie Otto for the "toxic green" label.

Pen-pals Michael Andrew Kreski, Ken Spencer, Hunter Collins, Hugh Foster and Ray Mahar. Do they not brighten my days enough to earn this vast dedication? Yes, they do; the whole page except for a tiny piece somewhere down in the corner.

In the early seventies, the British government was said to have been offering financial aid to developers willing to invest in its Turks & Caicos Islands, down there below the Bahamas, so some rather naively idealistic investors flew me down to those gorgeous and, at that time, largely unspoiled islands

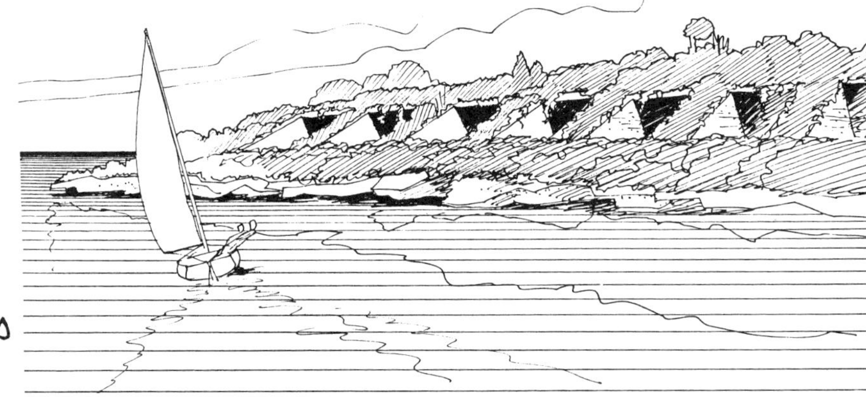

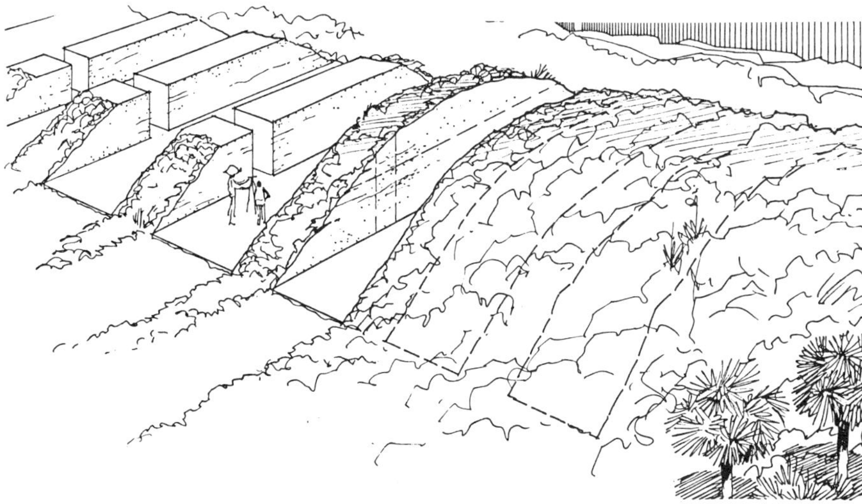

to see if simple, environmentally-benign projects would be encouraged. The answer from the authorities there being a qualified yes, I returned to New Jersey and worked out this carved-into-the-coral-rock design.

In the Turks and Caicos Islands, everything was scarce. Fresh water, wood, cement... even skilled labor... but the people of the area were extremely friendly, enthusiastic, and willing. All that was needed was a nod from the Crown. That nod, apparently, was a bit more difficult to get than was at first indicated. As the red tape piled up, the investors began to lose interest, and, before long, this dream, called Stone Angel Harbor (or perhaps Harbour, down there) evaporated, as had so many of the others.

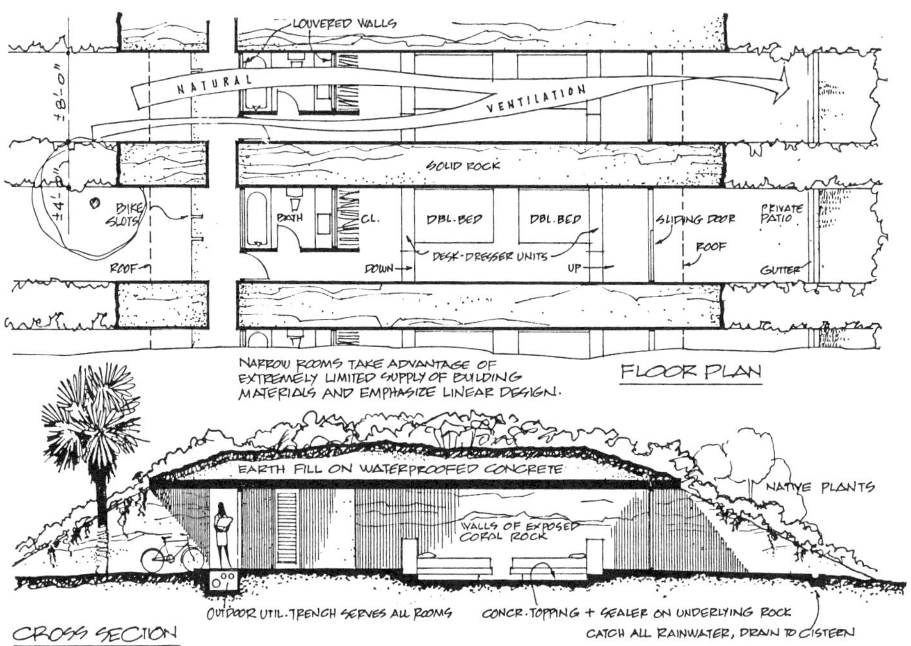

But what could have been simpler than carving 8-foot slots through a low ridge of coral rock? The natives had been cutting that stone for years. Add a roof slab, waterproofing, and soil, and you'd have guest rooms naturally cooled beneath the blazing tropical sun.

This idea was resurrected in a somewhat different form for the equally-ill-fated China project shown on pages 126 and 127. No doubt it will turn up again, somewhere, so be forewarned: if you come to me with a commission to design a hotel in a hilltop this is what you're likely to get.

It will, of course, be beautifully redesigned, but don't let that fool you.

I don't want to forget my favorite editors, none of whom I've met. Peter Fossel, Tom Rawls, and Chuck McCullagh. I want to make them immortal by airing their names here as I dedicate to them at least 8/9ths of every-thing visible on this page.

Every once in a while, I walk away from a client before he has a chance to walk away from me. Unless I'm mistaken, that was the case with this project for what was to have been called Ocean City Community College, in Maryland. The site was the all-too-familiar farmland about to be asphalted by suburbia, and, once I heard that the school would stress the notion of self-reliance, I produced a whole sheaf of drawings illustrating this sort of design. It was received with enthusiasm but funds were tight and the project began to drag. At a second meeting, I decided to see what the better-known part of Ocean City, Maryland, was like. I pictured a quaint seaside village something like the Ocean City, New Jersey, of my childhood. Instead, I found a wall of high-rises running for miles along the beach. It was a carbon copy of Miami Beach's Collins Avenue.

I never bothered this client again. That tortured ocean-front had done me in.

Three writers have tried to write about this underground architect. All went away too bored to do it. Brad Lemley was the exception. His writing is so good the subject doesn't matter, so to you, Brad, I dedicate all but 3 words printed on this page.

Cape Cod, on the other hand — back in 1974 — <u>was</u> like the New Jersey of the Thirties. I was amazed to find that the peninsula was a vast pine forest with a few small towns along one or the other of its saltwater sides. As soon as I saw it, my head packed up

and moved to the Cape, even though my body didn't make the move for another three years.

That first, 1974, trip was made for the purpose of designing this underground house for an Orleans family. At my suggestion, they had a local architect take over the design. (He improved it quite a bit.)

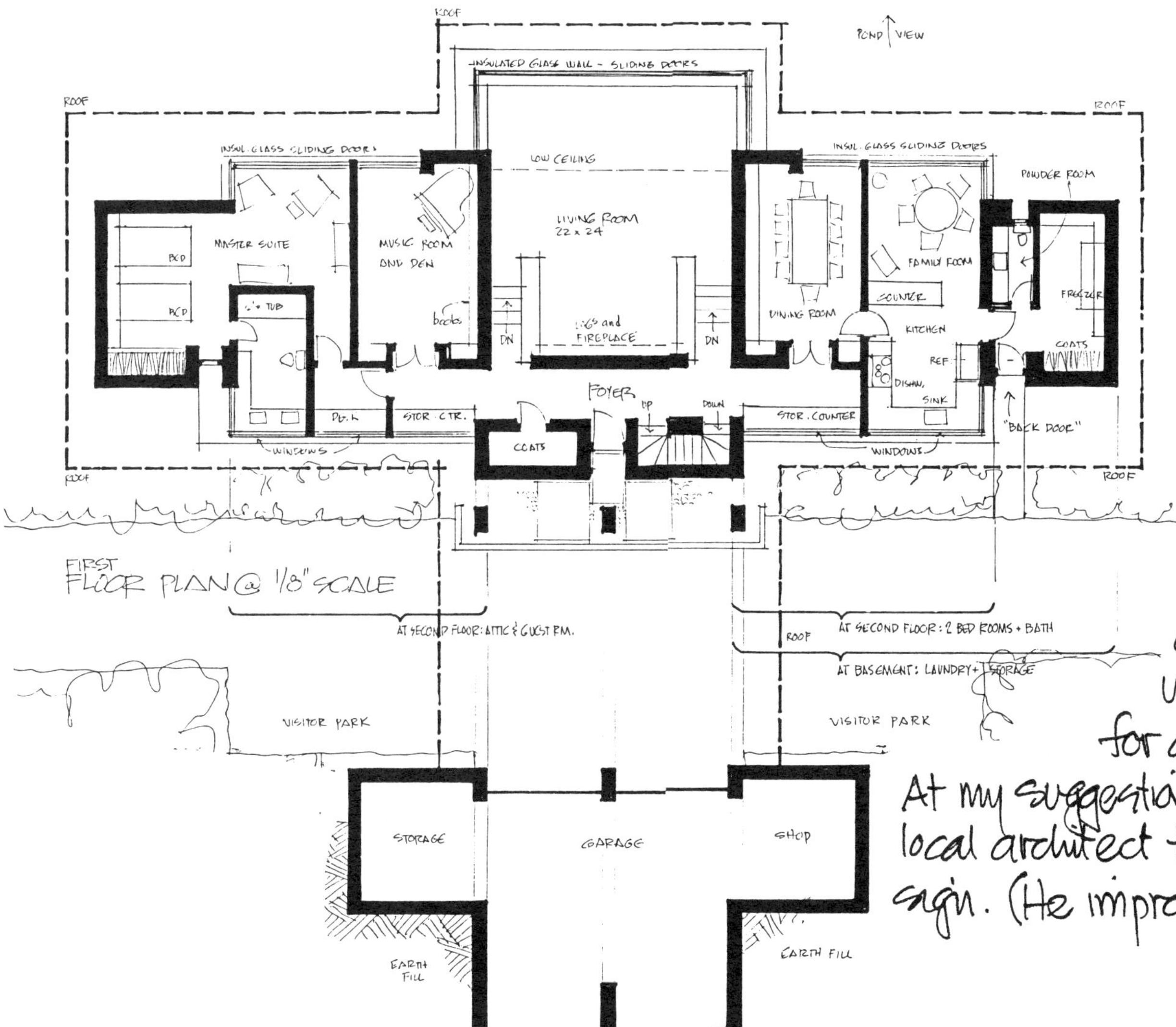

Another freind I've never met, Kate Siegel Bandos, has shown me how to get books into the hands of reviewers. She must have a magic key. So to her I dedicate 23/27ths of the ink and paper on the page you now peruse.

This house was based on an idea so simple that many versions of it have been built. The original one, the one called SOLARIA, below, was built at Indian Mills, New Jersey, in 1974. My friends Bob & Nancy Homan were willing to take a chance on several then-untried systems, the main ones being solar heating and earth cover. I'd never piled dirt on a roof with such a slope but after some trial and error (I don't want to talk about it) horizontal cleats were added just under the rubber waterproofing sheet, and the earthen fill stayed right where it was put. For a solar heating system we chose the super-simple trickle-down panels invented by Dr. Harry Thomason. Both the earth cover and the solar heating worked like charms, cutting energy bills to insignificance, and that's the way it went for 10 or 12 years. Then some

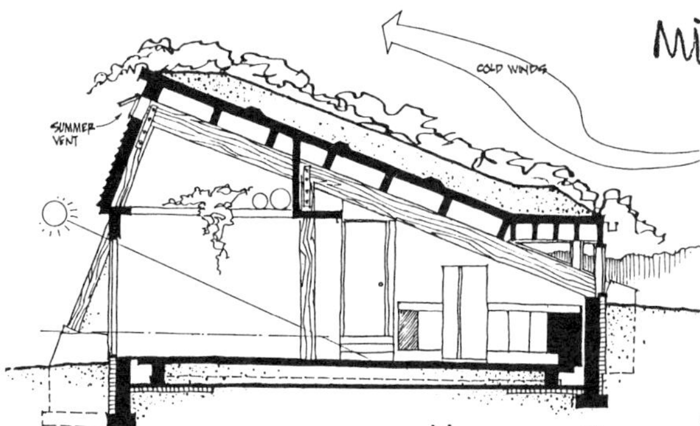

There's hardly room here to dedicate most of this page to all my friends at Raven Rocks, but that is still what I intend to do. To all of you Ohioans — and along Crum Road — 1445 — who've been such good friends, 14/17ths of this page is dedicated exclusively to you.

leaks began to appear. The Homans did everything to caulk and seal the openings as they were discovered but the problem kept growing. It was our first experience – and only experience, so far – in which thick, tough butyl rubber sheets admitted water.

After some awful messes and after trying everything else to solve the problem, the Homans bit the bullet and took all those tons of earth off of the roof. The butyl, they found, had been riddled by ... what? It took some time to deduce that beetle larva (that had, apparently, been in the big roof timbers from the start) had bored through insulation, plywood, rubber, and earth in order to reach the sunlight. And the Homans, having had enough of roof troubles for a while, put a conventional roof on Solaria for the time being.

The moral of the story could very well be don't build underground! Could be. I prefer to think that the advantages still far outweigh the disadvantages, and that we must all be prepared to accept the occasional hard knocks dealt us by fate as graciously and as good-humoredly as the Homans did.

My first underground design for them, by the way, before my solar conversion, looked like this. (⟶)

You'll see more of this one on the photo pages.

The Cary Arboretum of The New York Botanical Garden is in a lovely valley at Millbrook, New York, not far from Poughkeepsie. When the trustees of the Arboretum invited me to design their new Plant Science Building they not only offered to try some new ideas, they actually insisted upon it.

The result: earth shelter, solar heating, thermal mass, exterior insulation, task lighting, summer sunshading, water-conserving plumbing fixtures, heat recovery systems, and, last but not least, appropriately, a landscape of native plants.

Unfortunately, the solar heating system turned out to be unsatisfactory and was removed, delighting some anti-solar skeptics who assumed, incorrectly,

A special pen-pal and old friend is Wayne Stroup of Pendleton, S.C., who can form words so beautifully they're a joy to read. To you, Wayne, I want to ded-icate 1/14th of this page in the hope that you'll continue to send me more moving messages.

that the sun itself had somehow let us down.

Here's part of a letter I wrote to the owners of a monastery, also in New York state, who had invited me to join them for a couple of days before submitting a proposal. The robes and the meditation to which they exposed me never took root but the wholesome food and the silence, not to mention the genuine friendliness, really moved me.

The pages of design ideas I sent them seemed to move them in return. We stayed in pretty close contact for a while.

Maybe the phone will ring any day now.

Break the car-mood dramatically; abruptly; with a land-centered, rather than a man-centered, entrance to the community itself.

I see a long, gently-sloping up-ramp as the central trunk of the country center. From an entry/foyer, which might include some business offices as well as the usual shoe-removal, coat-hanging, and clothes-changing rooms under its low roof, one could pass through a series of ever-loftier, ever more awesome spaces toward the heart of the complex. The point, of course, would be to use architectural devices to put the jangling 1975 world behind, both physically and symbolically. The ceiling might start at a low seven feet, and slowly climb to greater heights as the space became narrower and narrower, the sky-glimpses fewer and fewer, until the space was nearly vertical; silent; shadowed... like this:

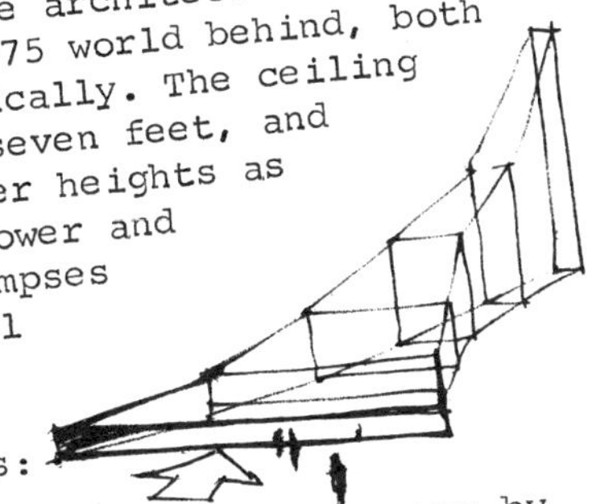

You might very easily create this kind of changing space-character by doing the work sequentially, in order, starting right at the entrance, using only untrained people working with the crudest of tools, and slowly letting the quality of the work improve as the workers gained experience, until, deeper inside the hill, nearer the entrances to the Zendo and the Buddha Hall, some very fine finishes might be achieved. I don't think you'd have to worry about the finishes becoming too perfect, not as long as human beings were involved.

Natural cooling, and even heating, using the earth mass around a building, is an idea that continues to generate a lot of interest. I was all set to install an earth-tube temperature-moderating system at my first Cape Cod house when I heard someone say that moisture condensing in such conduits could act as a breeding ground for Legionnaires' Disease. That did it: I haven't done an earth-tube system to this day, even though it is obviously no problem to provide condensate drains in the pipes.

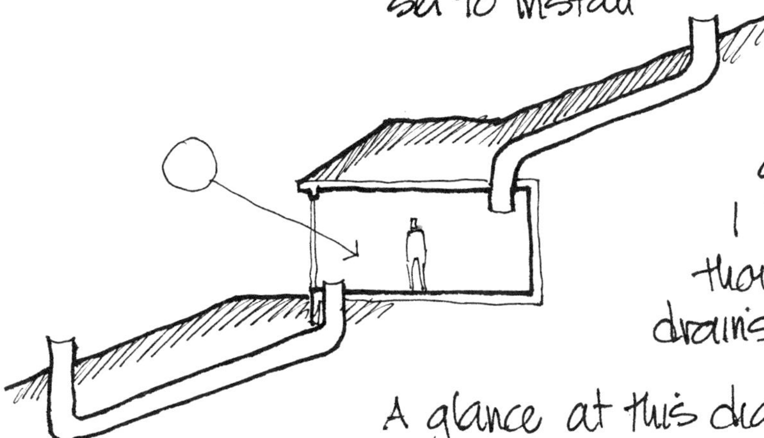

A glance at this diagram will make it obvious that cool air would flow naturally down through the building in hot weather, and that (relatively) warm air would rise through the building in the winter. That's practically the whole story on earth tubes.

John Hait,* who has developed earth covered houses that store enough summer heat to warm themselves all winter, uses earth tubes extensively, snaking them for, I believe, hundreds of feet through the soil before they enter the structures.

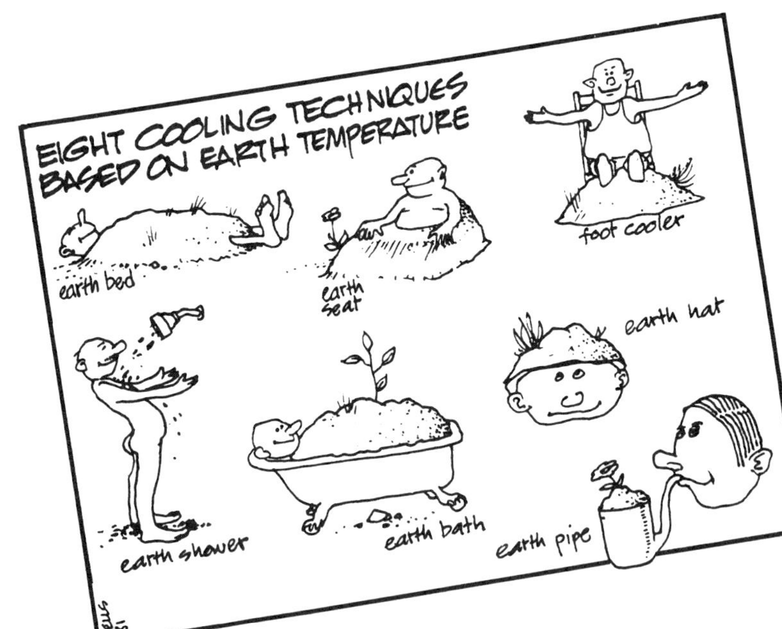

Laurie Virr was still a student when he came to work for me in 1963. Now he's a gifted architect in Australia. To him and to his wife, Mary, my old friends, I dedicate quite a bit of this page. It's another example of my generous nature, I suppose.

---

*Box 4694, Missoula, MT 59806.

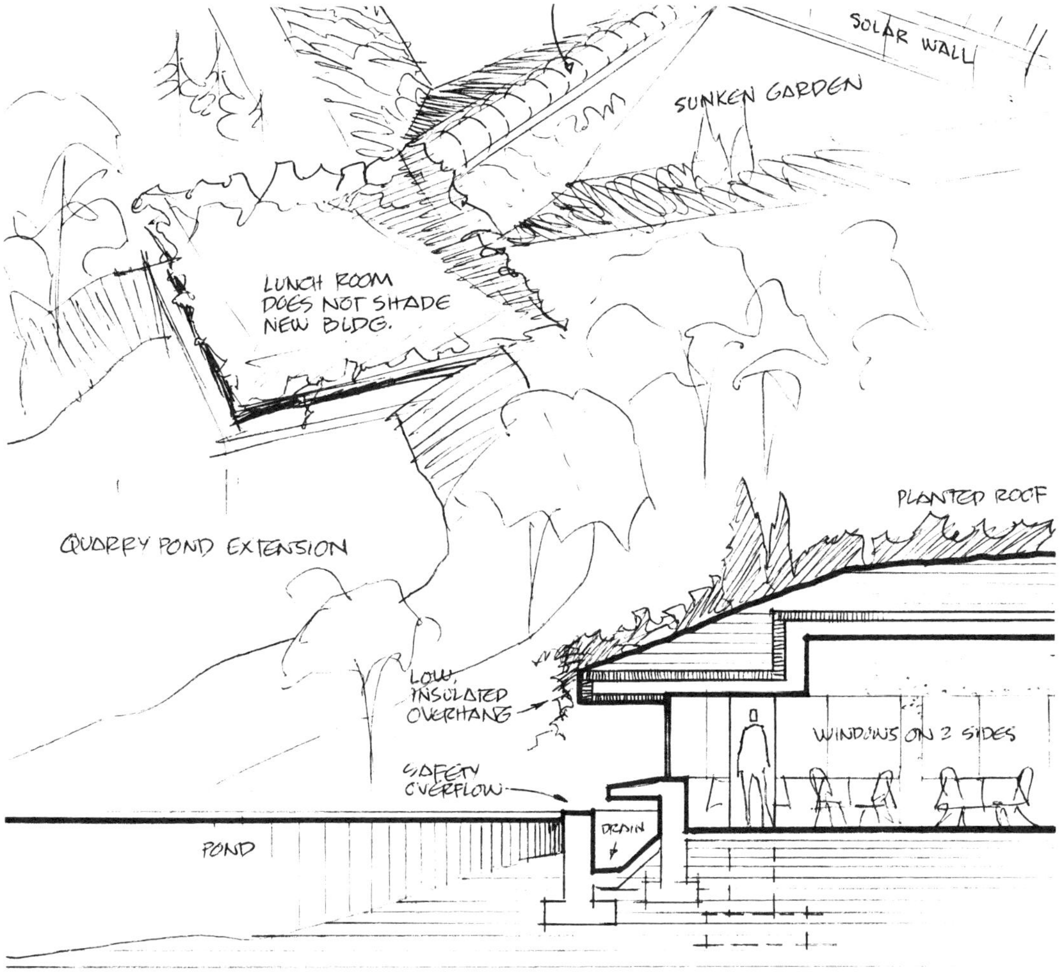

POND COULD BE MADE TO COME ALMOST UP TO WINDOW SILL (TABLE TOP) LEVEL.

You'll see this idea over and over again in these pages (pp. 104, 105; 140, 141) because I know it can be dramatic and trouble-free. A pond, either natural or artificial, is made to come right up to your window-sill, causing perhaps some initial uneasiness but after that, I think, endless fascination as ducks, geese, and sea monsters swim to within a few feet of you. This first attempt to sell the idea was made to a publisher in Pennsylvania who wanted to build an employees' (organic food, no less) cafeteria near a little pond on his property. I simply wanted to extend the pond a bit. If I were to do the pond-sill now I'd slope the top of the dam wall so that ice expansion would ride up

and over the top instead of toppling the wall.

And what about new friends? What about Stephanie and Frank Foster? Even though both of them shot me I lived to tell the story. This is your page—most of it. S. and F. Do with it what you will. This will probably make you famous.

During a telephone call from one of the more rapacious land developers in New Jersey, I found myself doing this doodle of architecture, greed, cruelty and sins of all kinds. This was in the mid-Seventies, sometime.

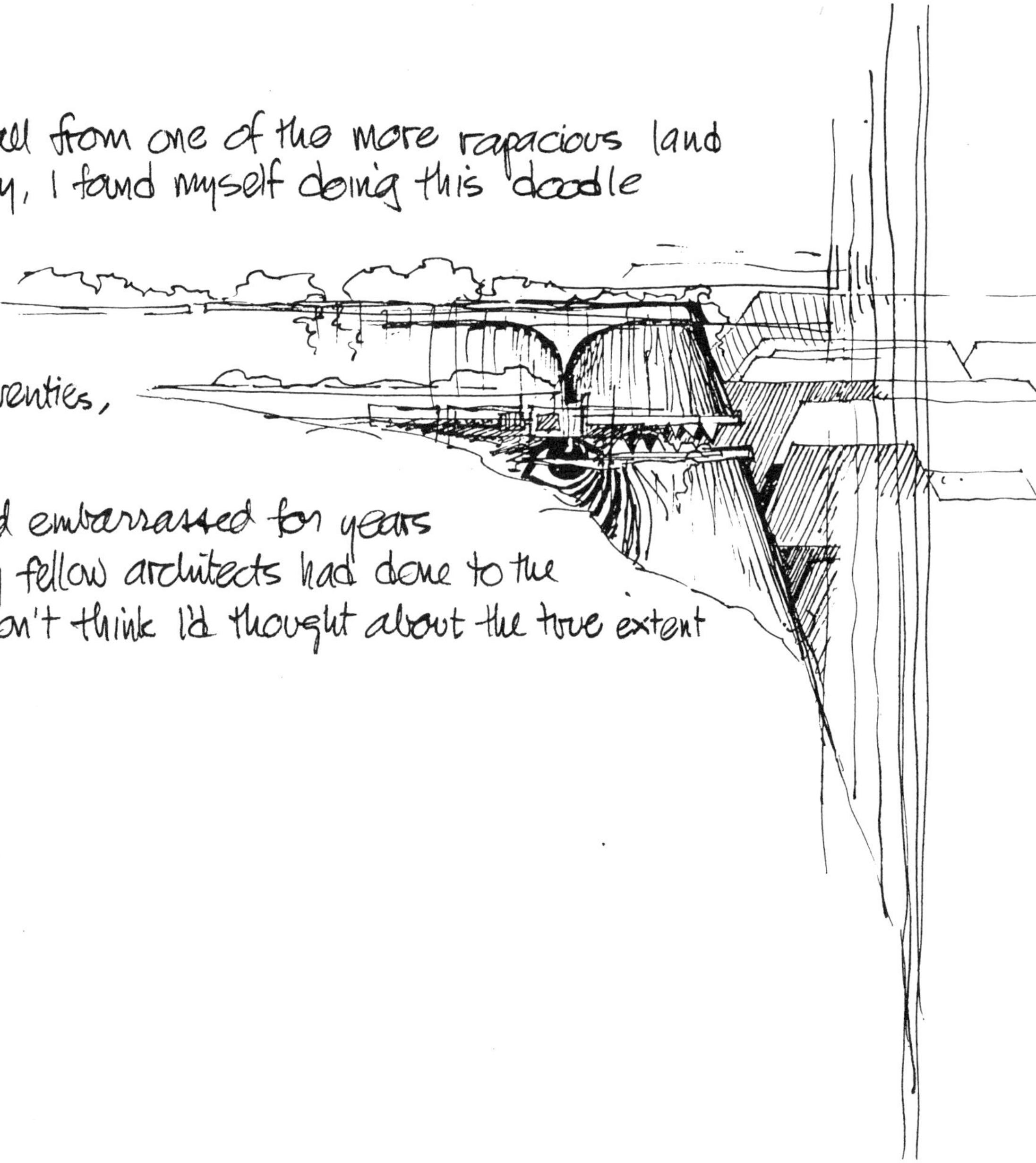

I'd been ashamed and embarrassed for years by what I and all my fellow architects had done to the American land but I don't think I'd thought about the true extent

Marie and John Schwalbe live in the very heart of bustling downtown Brewster. How can they stand that hectic life? I don't know but I want to dedicate 2/5 of a page to John, and 21/50ths to Marie. Good people, good friends.

of our roles in the scandal until I stopped to examine my telephone doodle. It had been aimed at the developer to whom I was speaking but it was obviously a picture of me. With my license from the state,

with my pencil and a t-square,
I'd be happy to "improve" any site for
you if you'd just pay me a fee.
Maybe I don't do ticky-tacky boxes
but my work is still so far from
the ideal I might as well be doing
them.

This cartoon, which grew out of the doodle at left, turned up in many places, surprising me in my travels to architectural schools.

All but 3/100ths of this page is dedicated to that very special person who, in my absent-minded-ness, I've failed to mention, even though you mean so much to me. This special catch-all dedication is for you.

Among the titles I've considered using on my underground books have been

IT'S ALL MY VAULT,

BEST CELLAR,

and this one, → seen here as part of a proposal I once sent to a publisher.

I want very badly to do a house that looks for all the world like a hill, and I want to build it so that all the floors, both indoors and out, are indeed on the level.

The title will have 4 levels of meaning,

1) Straight talk about the pros and cons of earth shelter,
2) Putting to rest the fear that you must go <u>down</u> into an underground building,
3) Full access by the handicapped, and
4) The potential for using storage carts (holding everything from cleaning equipment and home repair tools to little-used cookware and out-of-season clothes), making every part of the house easily reached and served. Tax records, party games ... a category for each cart.

Ronald North died in a Marine helicopter accident in 1989. Too young. He was my nephew. I should dedicate an extra page or two to him but I know he wouldn't want more than his share. So Ron, you get .91 of a page as my memorial to you.

The floors will of course be pitched as much as is required for proper drainage, and to allow for future settlement of the soil, but to anyone visiting them they will be essentially flat. Toddlers can toddle, oldsters can waddle, and all the wheeled devices in our lives will roll through the spaces with ease where in most buildings steps and curbs and too-narrow doors have even the strongest of us groaning as we carry big bundles, 5-gallon paint buckets, pianos, refrigerators, and 155-mm howitzers from room to room.

ALL-ON-ONE-LEVEL CONSTRUCTION IS NOT ONLY KIND TO THE HANDICAPPED, IT'S GREAT FOR USING THE DOZENS OF ROLLING DEVICES NOW AVAILABLE. (THE LOCAL WELDING SHOP MADE THE FRAME OF THIS FOR $20. I DROPPED IN 3 PLYWOOD SHELVES, ADDED CASTERS, USE THE CART ALL THE TIME. — M.W)

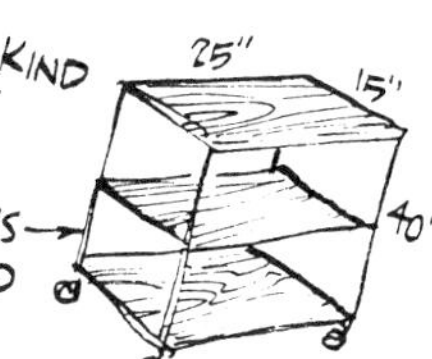

"You want on the level," said fate. "Here, I'll give you on the level. How's this?"

Someone in Ft. Worth with a lot of land had been moved by pictures of my work and, as a result, invited me to design a house out there – way out there. In true western-hospitality style I was sent first-class plane fare, visitor information books, and a hotel reservation!

To all my brothers- and sisters-in-law: greetings. Most of what you see on this entire page is dedicated entirely to you. It will bring you fame and happiness beyond your wildest dreams, so get ready to stand in the spotlight.

It was the royal treatment and I lapped it up.

I was met at the airport, wined, dined, and taken to my hotel. In my room I found rented-car keys, a map of the route to the site, anti-chigger lotion, and a snake-bite kit! And when I got to the site, there, believe it or not, was a picnic lunch in a cooler/hamper, waiting for me at the gate.

Now we've come to the publishers' page, but I no longer have a publisher. I publish these books myself, so a lot of this page is dedicated to Allen Bragdon, who is a publisher but he's here strictly as a friend. Ain't that so, Allen?

The owners were, among their many other interests, avid vegetable gardeners. They wanted what might be called a gentleman's farm, out there on the plains. They saw their land as a Ft. Worth suburb and, indeed, I could see the towers of the city on the horizon when I visited their land (keeping one eye out for the rattlers I knew were behind every blade of grass). The owners wanted to divide their holdings into a lot of mini-farms for people with similar interests, 10 or 20 acres to each site.

I just couldn't imagine putting their little farm complex out on the naked plain if it would make a nasty little bump on the land, so I proposed a sunken courtyard reached by driving gently down a long, curving driveway at such a gradual slope the earth banks along the way would seem to rise up out of the land until suddenly there was the courtyard, as lush and green as only that land can be when water is available. Everything was to be earth covered except for a great long living area in which a garden – with trees – and a pool would lie beneath a conventional, but skylighted roof.

Everything went along smoothly. The owners were enthusiastic. I began

With these moving words I hereby dedicate all but three letter a's on this page to Lois & Bill Racz for being such good friends and for having such good liberal views in these strangely conservative times. Hang on to those values!

to develop details, to work out a dimensioned floor plan, to think about getting a Texas architect to tackle the job of producing construction drawings, and then...

oh, no!

...oh yes. Circumstances combined to make the move untimely for the owners, and they had to cancel the project.

That won't stop me from showing the rest of it to you, however.

And that's my Texas story.

I told you, back on page 1, remember, that this was to be a book of failures, so you can't very well complain.

For a final look at this house be sure to visit the next 2 pages.

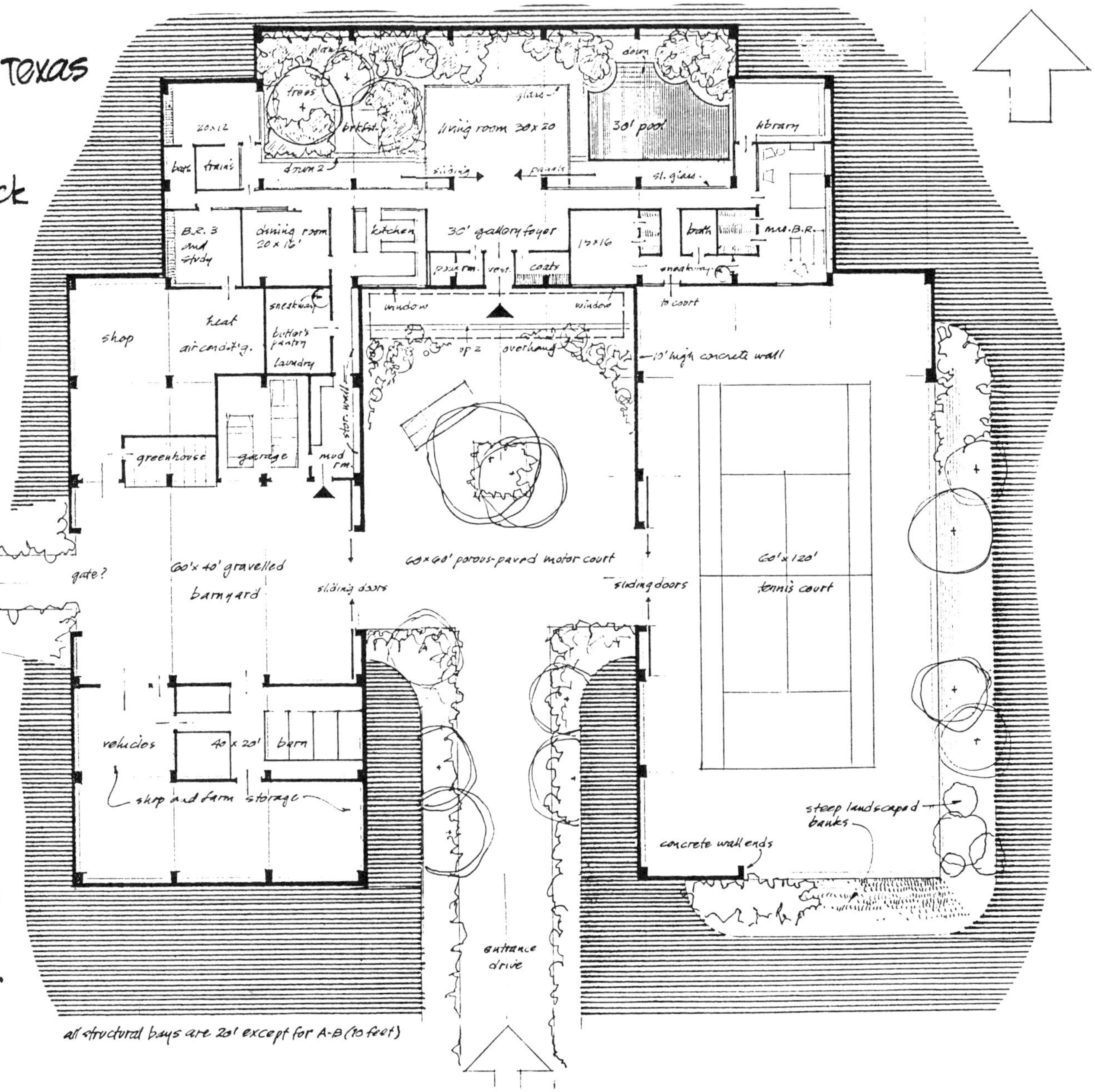

A special 9/11ths of a page is dedicated to all the environmentalists who work so hard to try to awaken all the rest of us so that we, in our work, will begin to reflect the ideas that have taken so very long to soak into our hides.

I've always wanted to bring a shingled roof right up out of the earth. Would it work? Of course it would work, easily, in a land not subject to hard freezes. The shingles, made of ceramic tile, would be set in mastic, perhaps, on the waterproofing, and then, as they came up out of the earth, they could be nailed conventionally to the roof deck.

Would the house itself have worked? Of course. The only problem I could foresee would be that of excessive humidity in the living/garden/pool area. But, with everything else behind glass partitions, the tropical climate of the great room would be no harder to manage than a room in Florida.

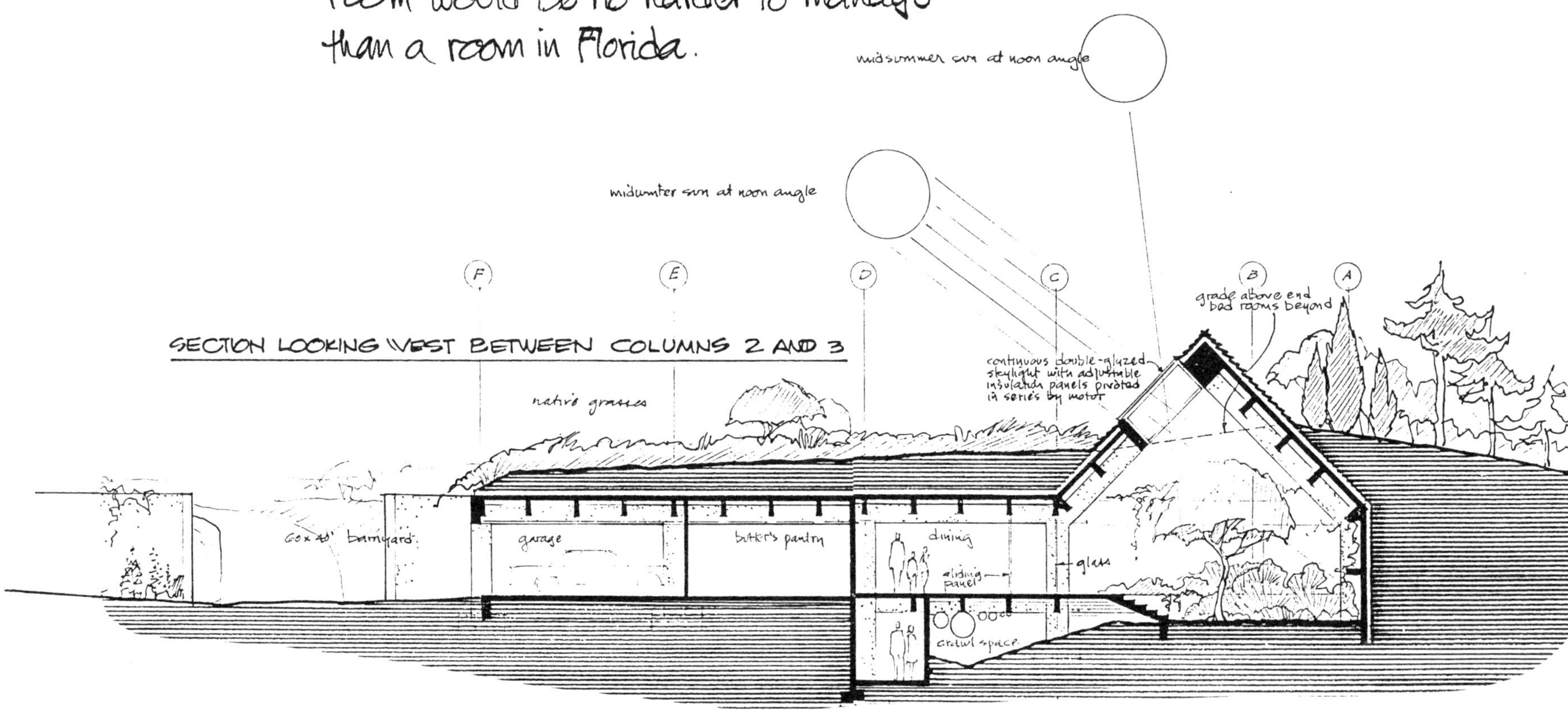

And now, if you'll forgive another abrupt change of subject, we come to the little house that, with a few modifications, can be entered from the south side as well as (as is shown here) from the north. This plan originated in 1977, and for 12 years it marinated in my head until, in 1988, it was ready to appear in POPULAR SCIENCE magazine as the "Do It Your Way Earth Shelter".

For details of that design be sure not to miss page 185. 184, too.

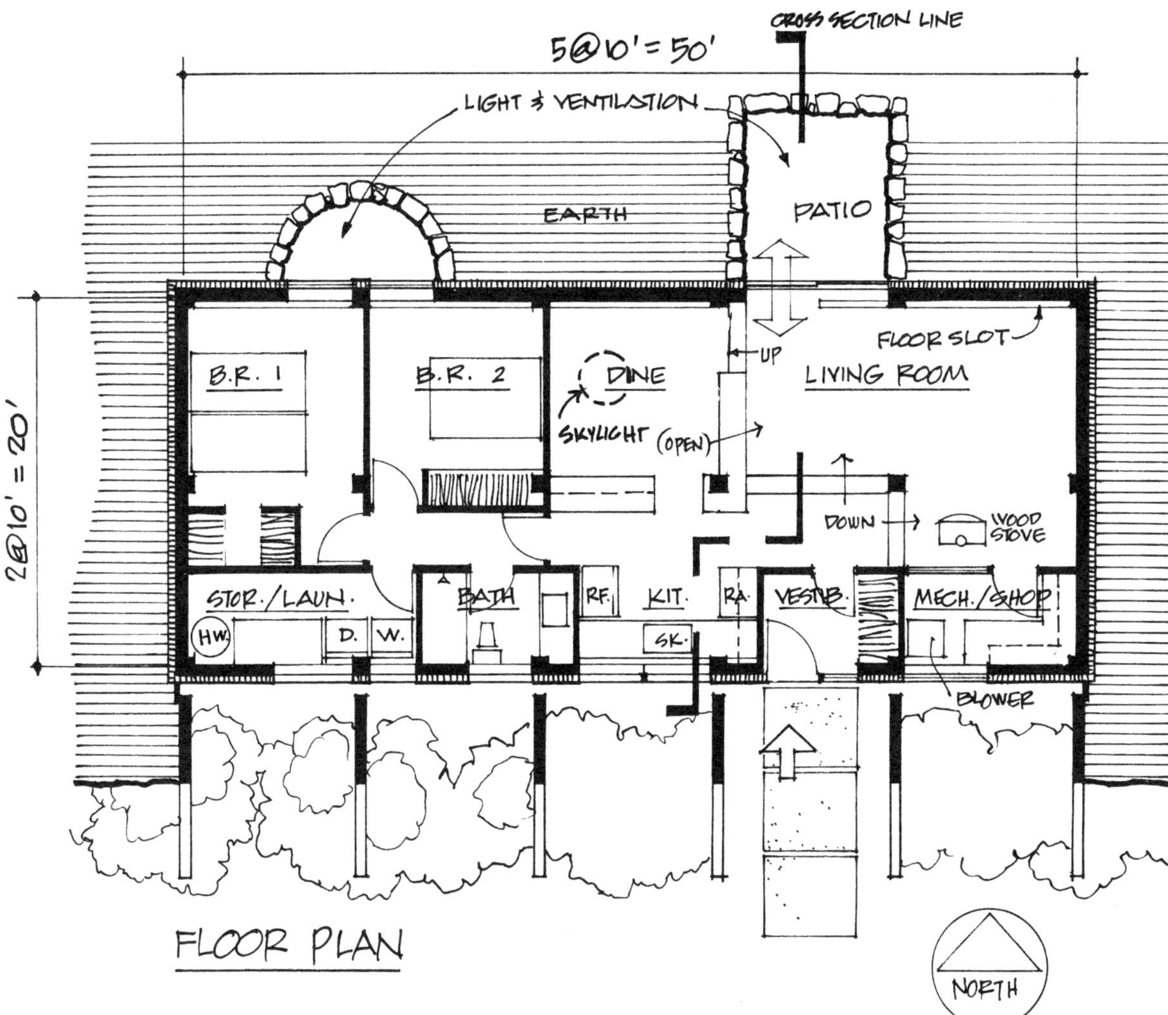

FLOOR PLAN

NORTH SIDE

Mid-Eighties: still working on it. The different versions get to look so much alike it's hard to keep them straight. Is this, for instance, the north side of the north entrance version? And does it go with the drawings below? Or is north the same as south in Australia at this time of year?

Something like that.

Natural cross ventilation, daylighted both sides.

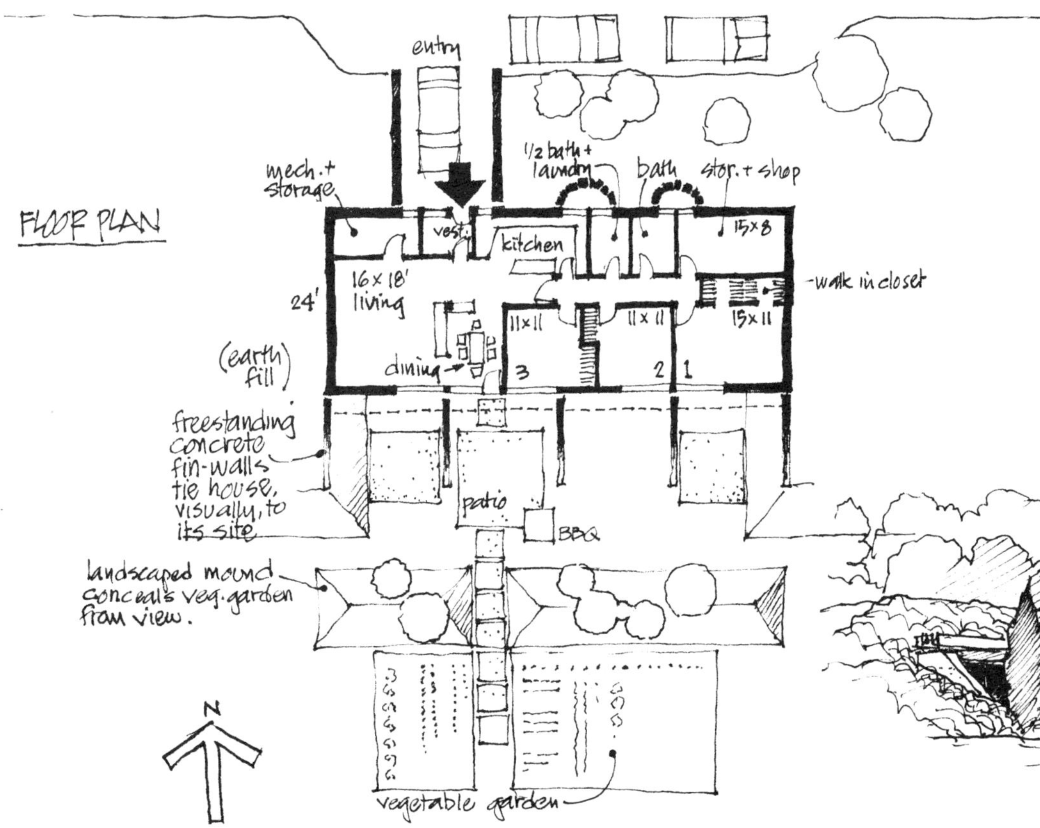

SOUTH SIDE

Well, then this must be the south side of the south entrance house. And this is its cross section.

THE GOAL: THE HOUSE AS PART OF THE NATURAL WORLD.

So the section should be on page 62 because... no, it's the south entrance house, too. See?

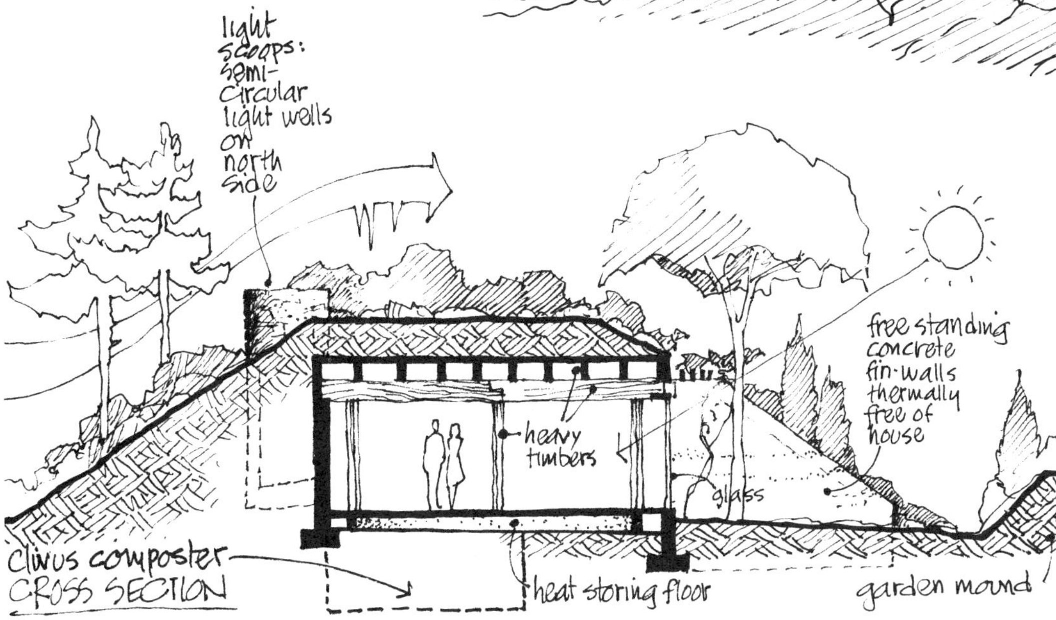

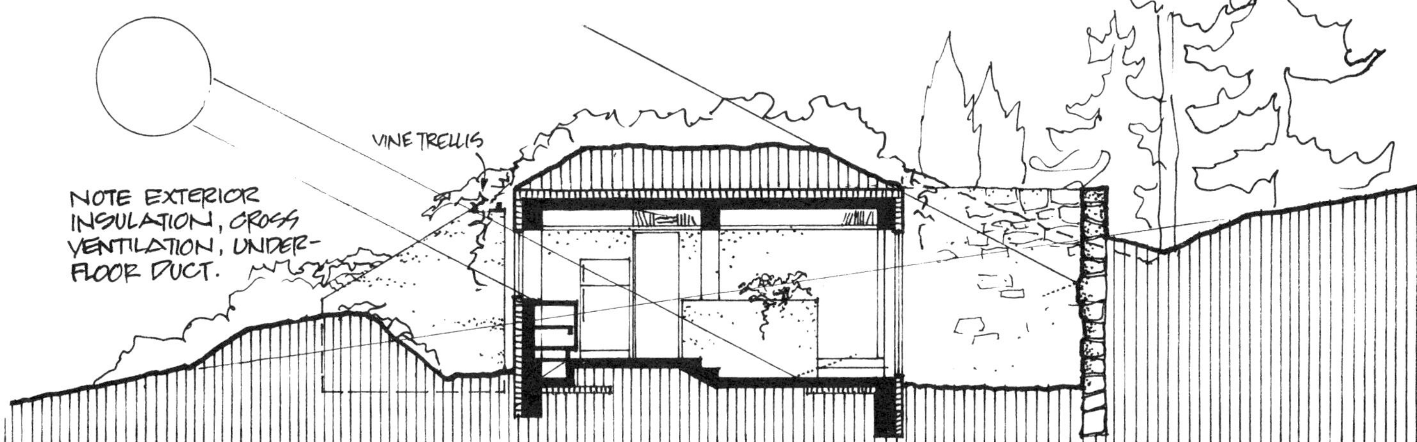

What would we do without good people like Phil Murphy who do things beyond repayment for us? We'd be in trouble, that's what. So I want to dedicate all but one line of this page to that quiet competent man who does not, apparently, keep score.

↑
plenty of glass on the south side
small windows or doors on the north ...
↓

That's what I was trying to say all along. Sun in, cold out.

The perspective above was done for some prospective clients who didn't think it possible to divide glass areas into large and small panes successfully. I won the battle but lost the war.

All the bits and pieces of pages left over from the dedications have been collected here as a special prize to be awarded to anyone who sends me ten or twenty thousand dollars in cash to help reduce the mortgages. Think about it, everyone.

By now, the meaning of just about everything on this cross section is no doubt evident but it does contain one element worth mentioning: deep excavation.

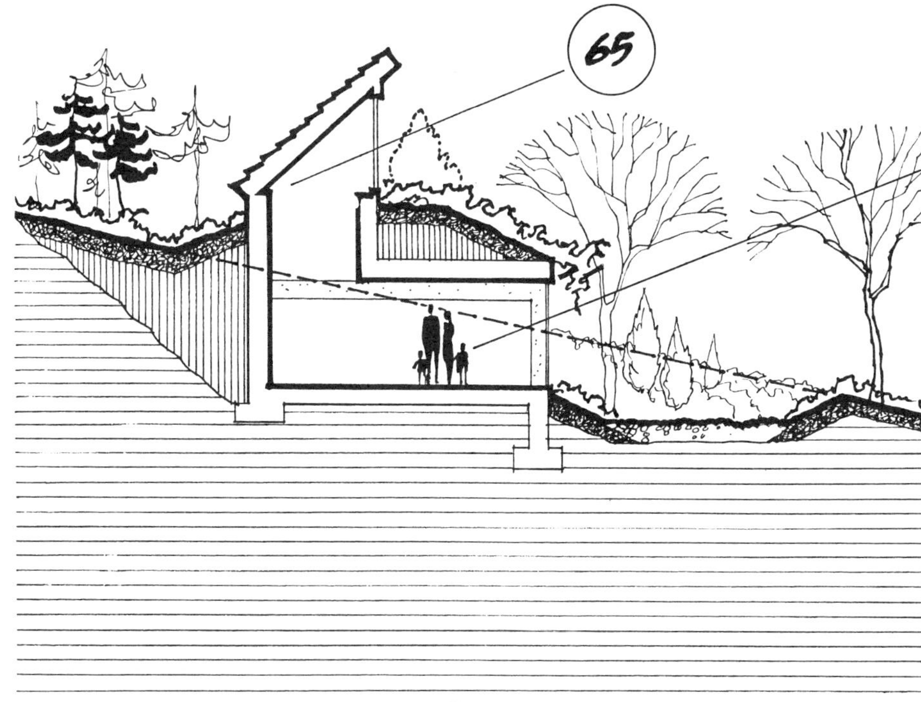

The vertical shading indicates backfill. Its depth on the north side of the structure suggests the great depth of cut required for underground buildings, particularly for those built on hillsides. The related problems are 1) the possibility of cave-ins, 2) the need to remove existing vegetation far to the north of the wall, 3) the danger of causing structural damage when replacing those hundreds of tons of fill, 4) the build-up of water in the relatively loose and porous backfill, and 5) the settlement of the backfill, even after it has been carefully tamped in thin layers.

Backfill, as it settles, or sinks, takes down with it such things as insulation board and (sometimes) waterproofing. It can also rupture buried utility lines!

The watchwords: be careful!

Here we see some more squares masquerading as triangles. Opening wide for the sun makes a lot of sense, especially in central Minnesota, where this riverside house was to have been built. Only by tunneling through a low,

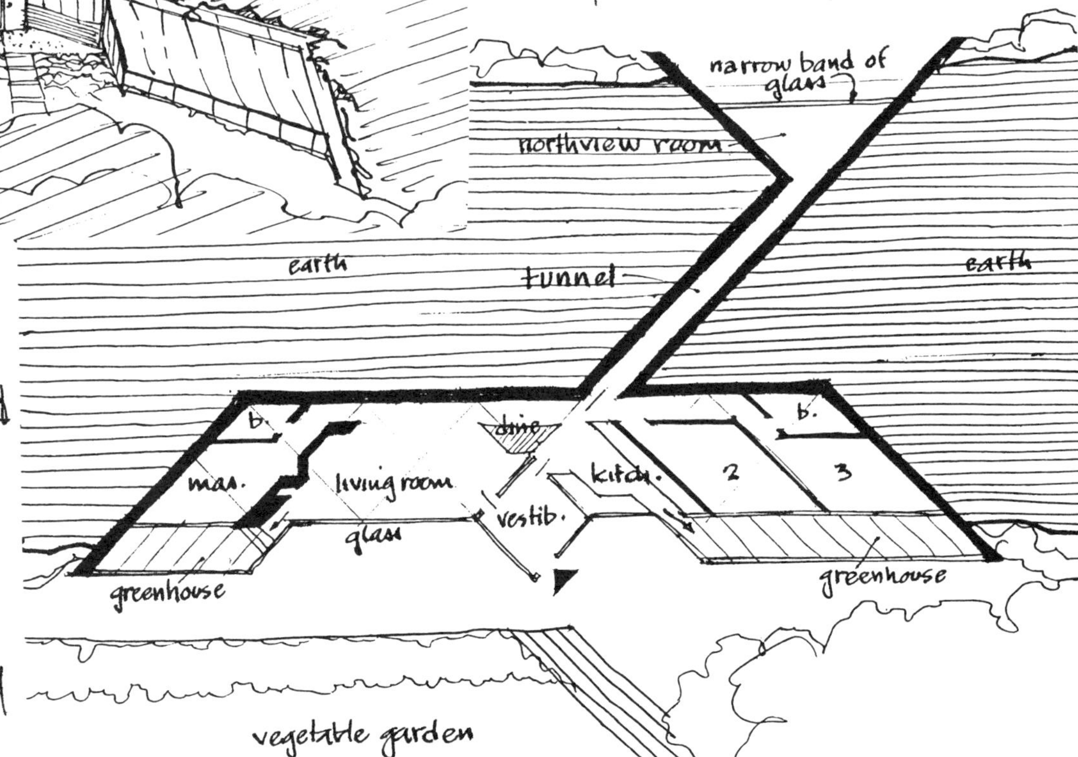

east-west ridge could the owner hope to see the river upstream. She was all set to build but then she fell in love with a vagabond and hit the road for parts unknown.

Architects love stairways – which are absolute barriers to the handicapped – and chimneys – which are notorious heat-wasters, especially when hitched to fireplaces. We love our fires, though, and let them go only with reluctance.

Could occasional open fires crackling on the hearth be forgiven by the goddess of environment if we did a lot of other things right?

There are times when it seems that an hour or so in front of a cheery blaze is like money in the bank ... the spiritual bank, that is. But if you say that fires have to go, I'll bow to your judgment.

How about a <u>half</u> hour?

The New Ponte Vecchio at Florence

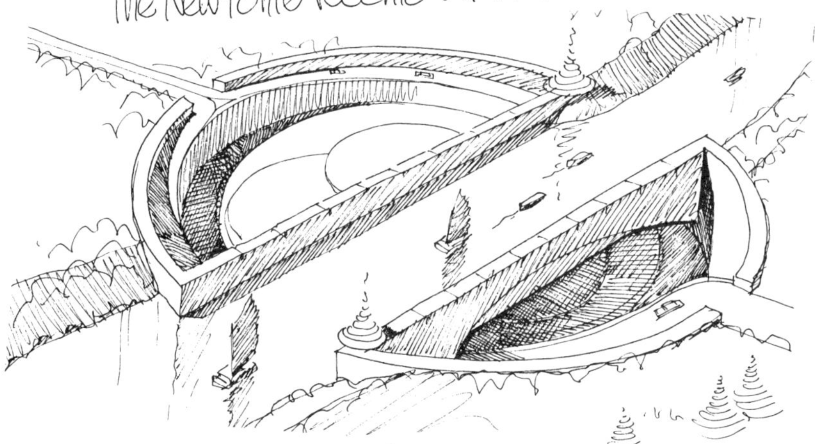

The Bridge Under The River Kwai

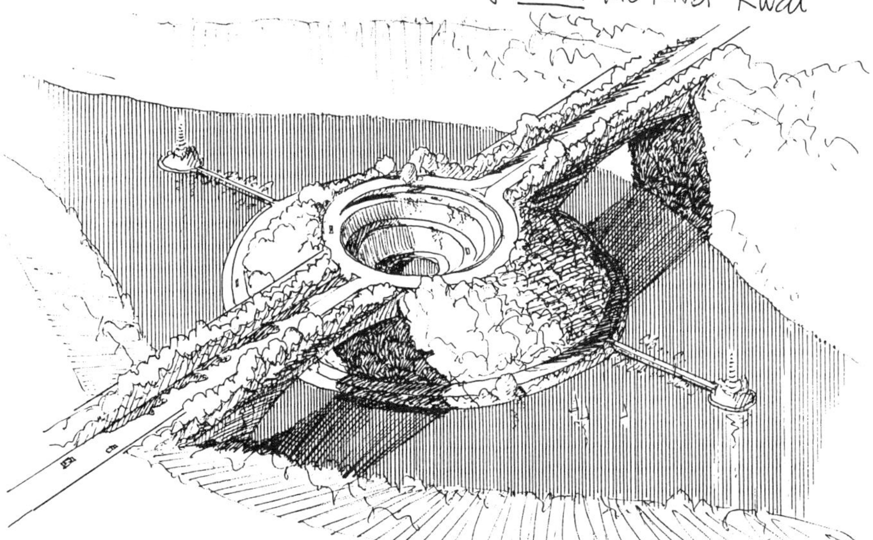

The Volcano Monument Cultural Bridge at Oahu

Fantasy time.

The little cities of Fargo, North Dakota, and Moorhead, Minnesota, as you know, face each other across the Red River of The North. Back in the seventies, a joint announcement was made: they would invite submissions in a competition for the design of a new bridge linking the cities.

I offered them several suggestions, of which these were only a few, but they failed to see genius when it was right in front of them, awarding the prize instead to an architect named Michael Graves. Well, they had their chance – and blew it.

Last I heard, the bridge had not been built.

Fantasy time II: Do you think that all the genetic manipulation now going on among the mad scientists of the world's research labs will ever give us the ability to grow our buildings organically? It's not inconceivable that such things could happen. I'm leery of genetic manipulation in general. There are bound to be disasters along the way, when the wrong kinds of organisms get out of the labs, but if we could ever learn restraint...!

I like the idea of watching houses emerge, develop, and mature, perhaps matching our own development and maturity. Recycling them would be no problem; their death and decomposition could match ours, too, so that the next generation of houses would have some good human compost in their make up.

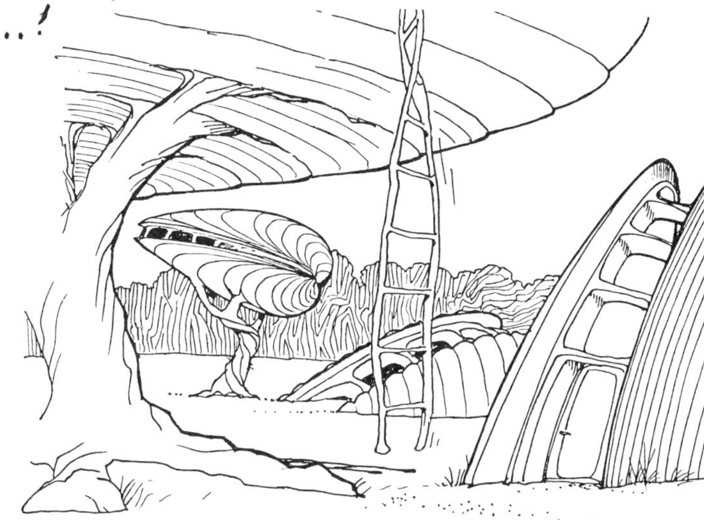

Talk about recycling! I didn't realize till I compared this design, just now, with the one on the next page how closely related they are. This was a proposal for a house at Wellfleet, here on Cape Cod. The road was at the top. A beautiful tidal marsh wasn't far from the bottom. A narrow step-down house was the result.

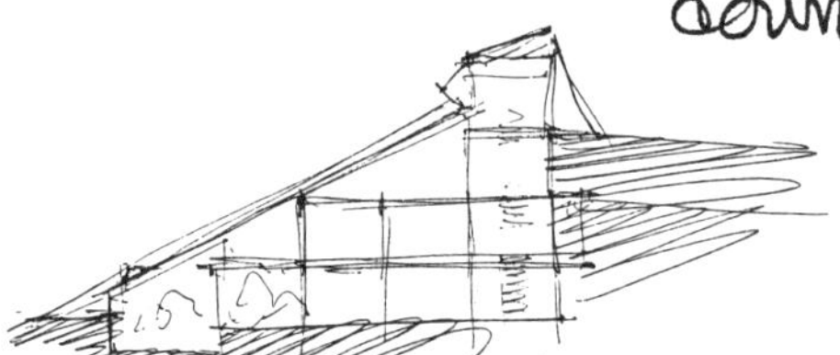

With the right kind of summertime shading and the right amount of ventilation it might have been both pleasant and comfortable. (In the winter it might have been condensate city.) And then there was the problem of hail, the threat of falling branches, and the annoyance of having little boys crashing through the roof.

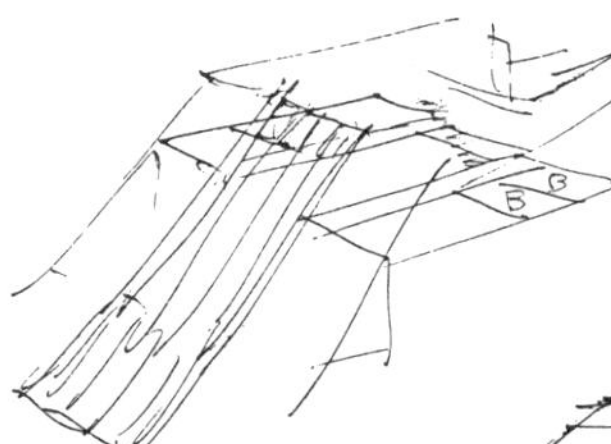

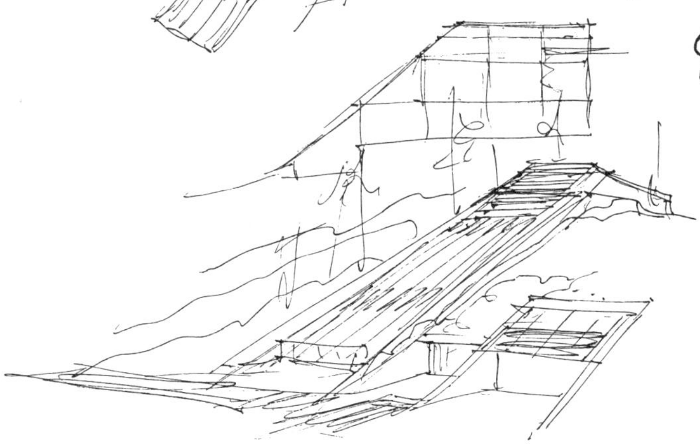

No wonder the owners tiptoed silently away.

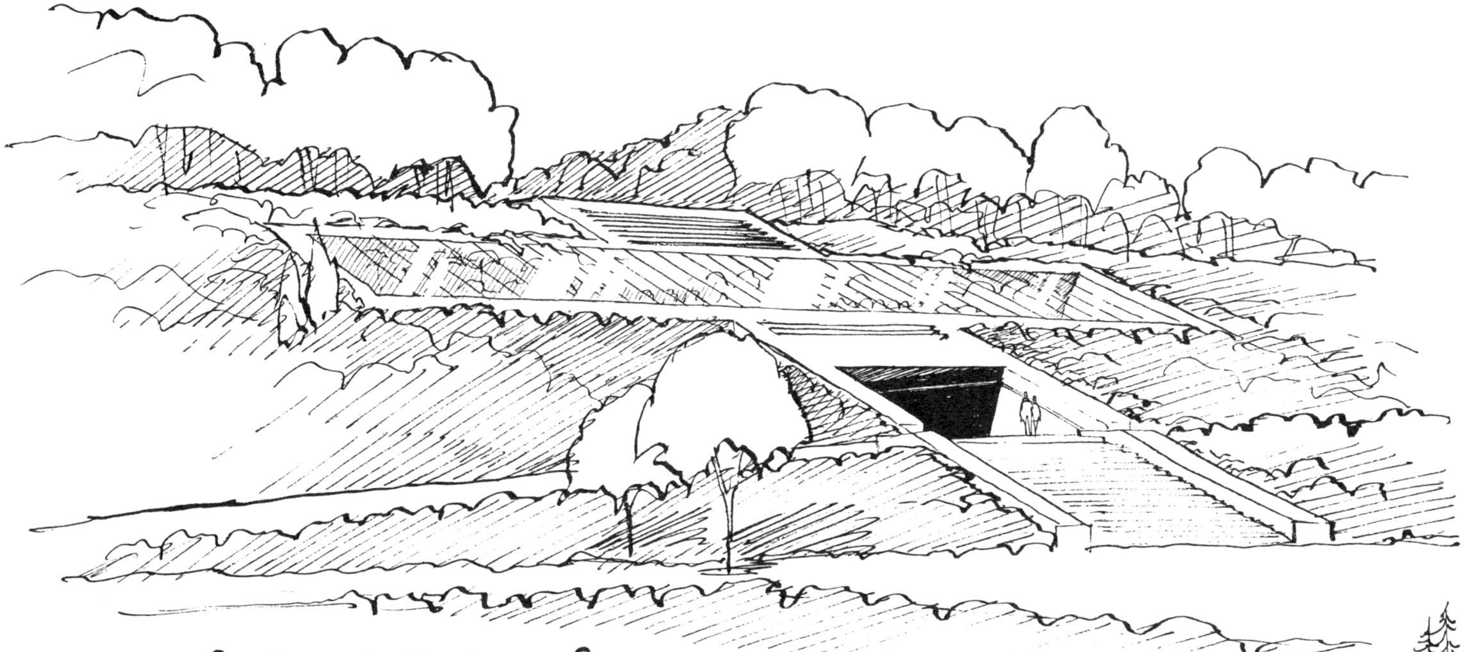

Wellfleet revisited; a few years later I proposed this design for a dormitory at a Catholic university in Minnesota. The great cross was not done consciously. 1978, I think it was.

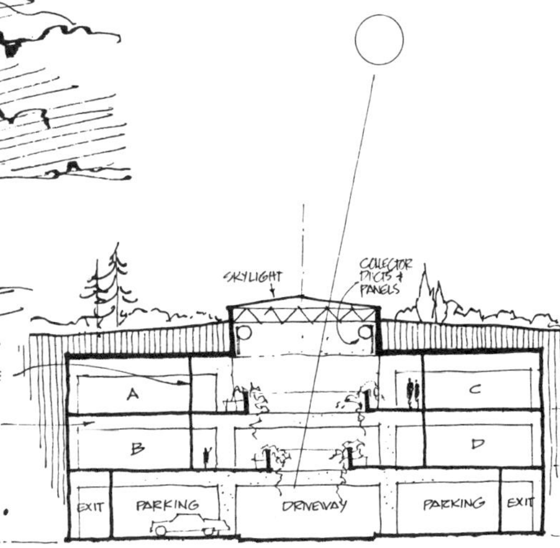

PLAN, @ '40" SCALE ... UPPER FLOOR

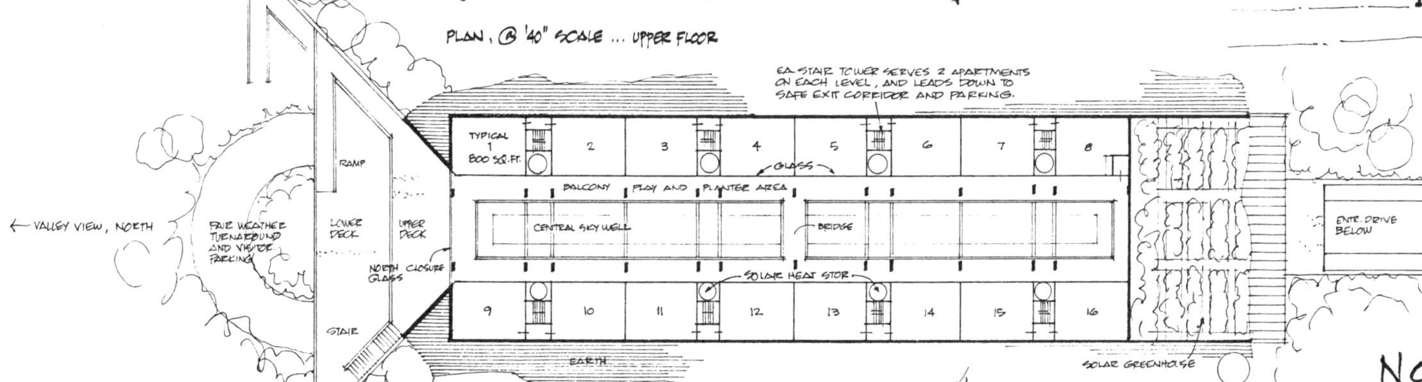

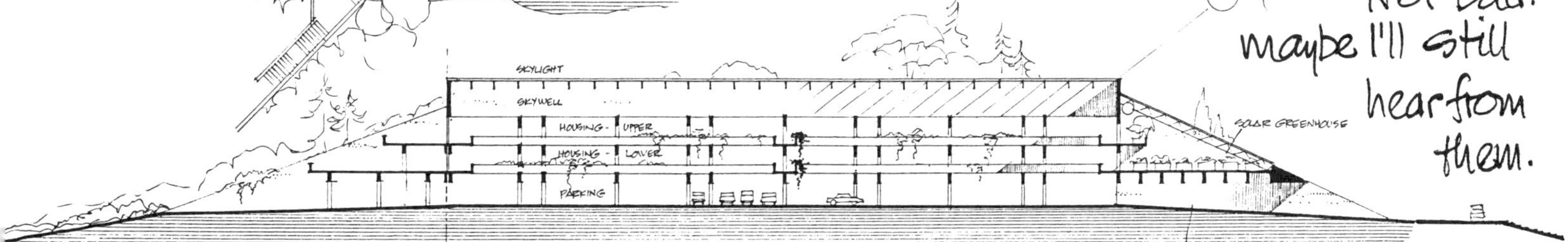

Not bad. maybe I'll still hear from them.

CROSS SECTION, LOOKING EAST, ALONG N-S CENTERLINE, @ '40" SCALE

Henry Ford's home in Dearborn, Michigan, was on an estate called Fairlane. The estate is now the Dearborn campus of the University of Michigan.

In 1979, when it was time to build a new environmental interpretation center, the university people reached all the way to Cape Cod for help.

After touring the campus and studying all the drawings they gave me, I proposed a building that sat astride a new pedestrian way leading from the heart of the campus to the woods along the river. To get to them, the main campus road had to be crossed, and the grades favored a pedestrian tunnel beneath that road.

This is a view from what I suppose you'd call the back of the proposed building through the tunnel and into the woods.

If I may refer you now to the site plan at right you'll see how it was arranged, and how another campus road was to have been diverted around the west side of the new

building. The main entrance, shown here (and in an earlier design on the next page) was to have been entered from that west (campus) side.

Earth cover, natural plantings, solar heat, task lighting... even an organic vegetable garden... all were combined in the hope that the building would not only do good things but say them.

THE UNIVERSITY OF MICHIGAN - DEARBORN
Environmental Interpretation Center

architecturally, as well. Ah, 'twas a lovely dream that slowly faded, as did so many of the rest, as funds dried up and the controllers of the moneybags turned to what they saw as more pressing needs.

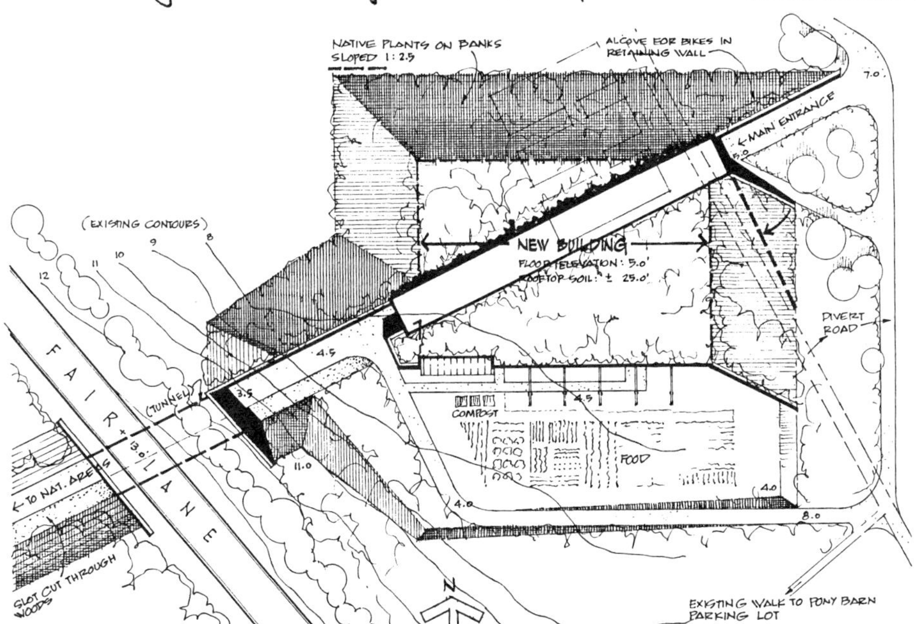

THE UNIVERSITY OF MICHIGAN - DEARBORN
Environmental Interpretation Center

(● = 7-11-79 REVISIONS)

0 50 100 150 200 FEET

PARTIAL SITE PLAN II

Here's that original west entrance design now, just in time to show us its potential for greatness as well as its tendency toward clutter, and its lack of shelter at the entrance doors.

As I recall, I failed to propose that the paved surfaces be paved with open blocks of precast concrete – porous pavers. They let grass grow and rainwater percolate where bikes and people – and even cars – travel.

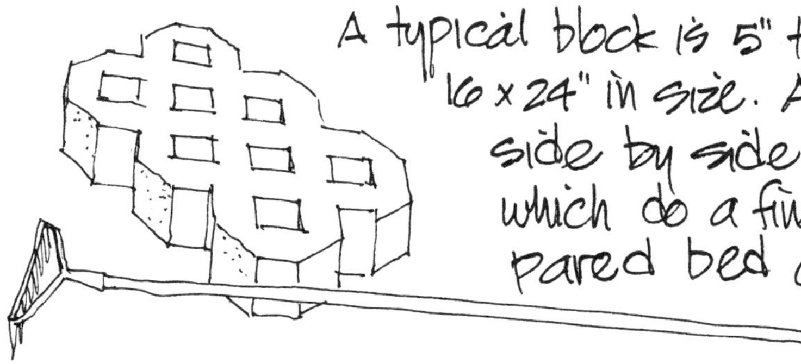

A typical block is 5" thick, weighs about 80 pounds, and is 16 x 24" in size. As you can see, the blocks, when laid side by side, form continuous rows of planted holes which do a fine job when placed on a carefully-prepared bed of crushed stone. Even snow-plowing is easy if the blocks are aligned.

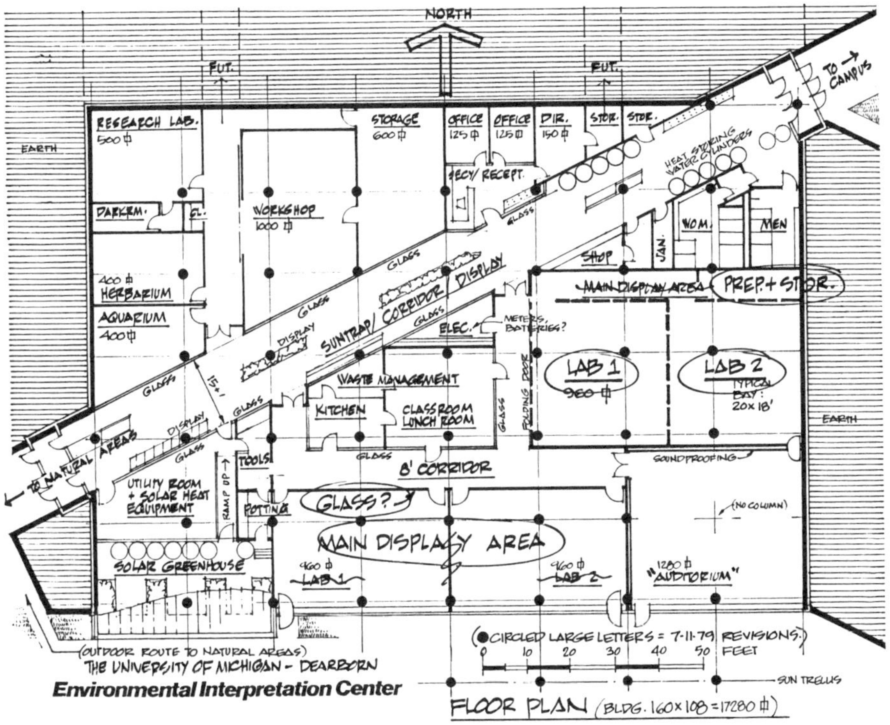

I'm afraid the only floor plan I can show you is this one with revisions marked on it. At least it will show you the campus-to-natural-areas axis as well as the use of insulation outside the concrete walls of the structure.

This lower plan is a daylighting diagram which, I notice now, would have left the Interpretation Center's director in the dark. Let's revise that right now before anyone else notices. All the other shaded areas needed, or could live with, artificial lighting.

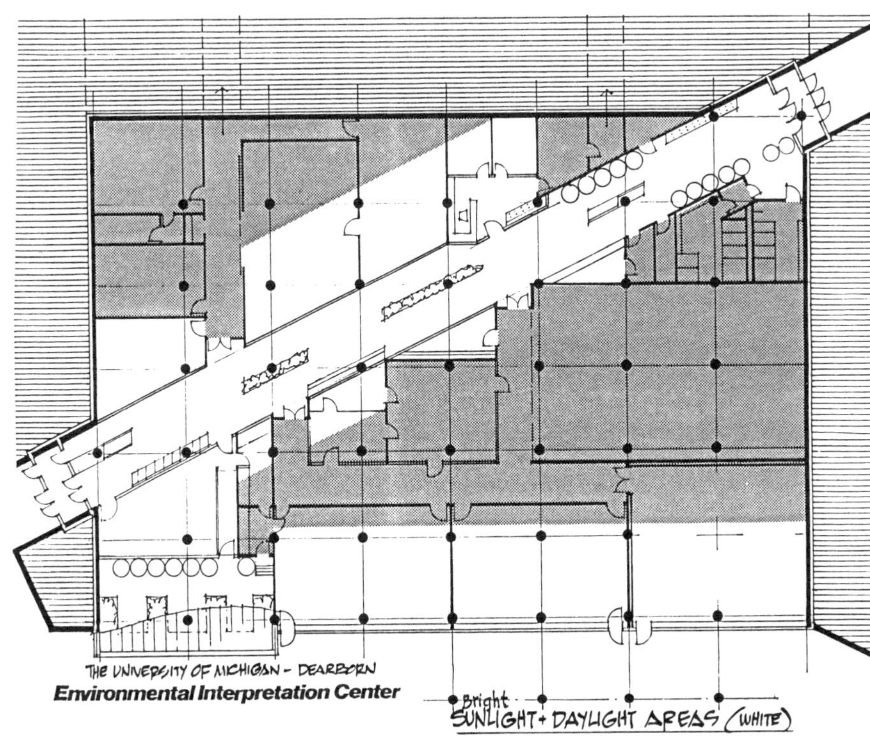

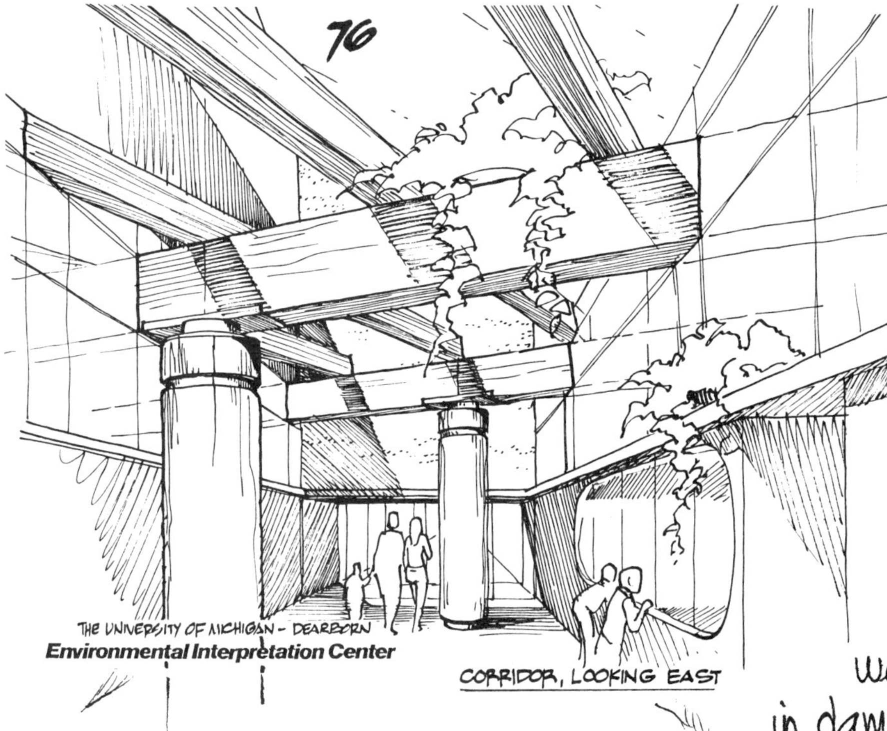

CORRIDOR, LOOKING EAST

The idea was to use a frame of giant timbers. There were several reasons: Everyone loves to see wood. Wood suggests the idea of renewable resources. And the huge columns would perhaps remind those who saw, and touched, them of the living treasures we so casually condemn each time we specify the use of wood. A 2x4, or a sheet of plywood, down at the lumber yard, is so unlike the leafy creatures we love to see in the forest we often forget its source and forget the cost, in damage to our west-coast woodlands, that results from all our lumber purchases.

These columns were not to be west coast imports, however, but rather the trunks of local trees cleared away to make room for the ever-spreading scourge of surface construction.

If underground architecture ever gets to the stage at which it could be called

THE SCALE OF THE SPACES

ever-spreading, it will be hard to call it a scourge as well. All construction causes land damage but underground architecture can heal the wounds, and in many cases can improve the health of the land, by getting the recuperation off to a good start and then standing back and letting the land work its wonders without further molestation, if you'll forgive the expression.

THE UNIVERSITY OF MICHIGAN - DEARBORN
Environmental Interpretation Center

CORRIDOR WINDOWS ABOVE ROOFTOP PLANTING.
AIR-TYPE SOLAR COLLECTORS ABOVE WINDOWS BELOW.
SOUTH/SOLAR FACADE

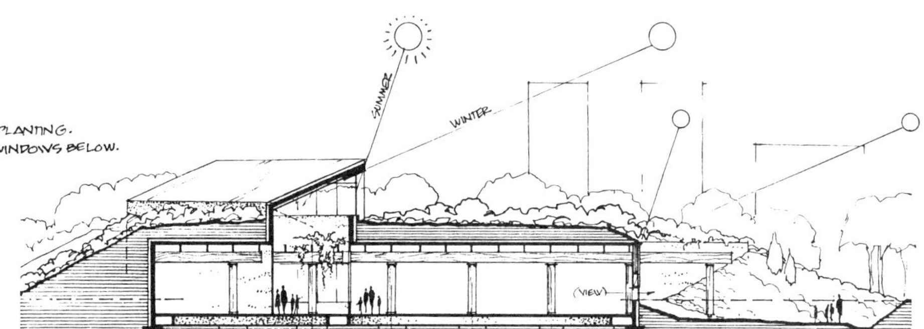

And so, as they say in the travelogues, as the sun sinks slowly in the west, we sail away from this lovely island of dreams to the next few islands, which, as it turned out, were made of dreams as well.

THE UNIVERSITY OF MICHIGAN - DEARBORN
Environmental Interpretation Center

DIAGONAL DISPLAY CORRIDOR
HAS 20'+ HEIGHT WITH CONTINUOUS 10' HIGH WINDOWS (INSULATED AT NIGHT) FOR SOLAR GAIN, CENTER-OF-BUILDING DAYLIGHTING, AND SUMMER VENTILATION.
SUN-WARMED AIR FROM COLLECTORS ON SOUTH FACADE, FROM GREENHOUSE, AND FROM HIGH CORRIDOR SPACE IS DRAWN DOWN AND BLOWN THROUGH ROCK BED BELOW FLOOR, FOR HEAT STORAGE. WATER CYLINDERS IN GREENHOUSE AND CORRIDOR STORE ADDITIONAL HEAT.
ELEVEN-FOOT-TALL WOOD COLUMNS, ABOUT 30" IN DIAMETER, SUPPORT MASSIVE TIMBERS CARRYING ROOFTOP SOIL.
ENTIRE STRUCTURE IS WATERPROOFED AND WRAPPED WITH EXTERIOR INSULATION BOARD.
TIMBER STRUCTURE EXTENDS BEYOND SOUTH FACADE (RIGHT) TO FORM SUMMER SUN SHADE.
NATIVE PLANTS IN DEEP MULCH COVER ROOF, SLOPES, AND MUCH OF SITE. FENCES HIDDEN IN LANDSCAPING PREVENT FALLS FROM ROOF EDGES.

CROSS SECTION LOOKING EAST

This plan, too, is just another square one, masquerading as an angled design.

I love excuses to walk out of doors, and these clients said they did, too. That's why the carport was to have been reached by way of a 50-

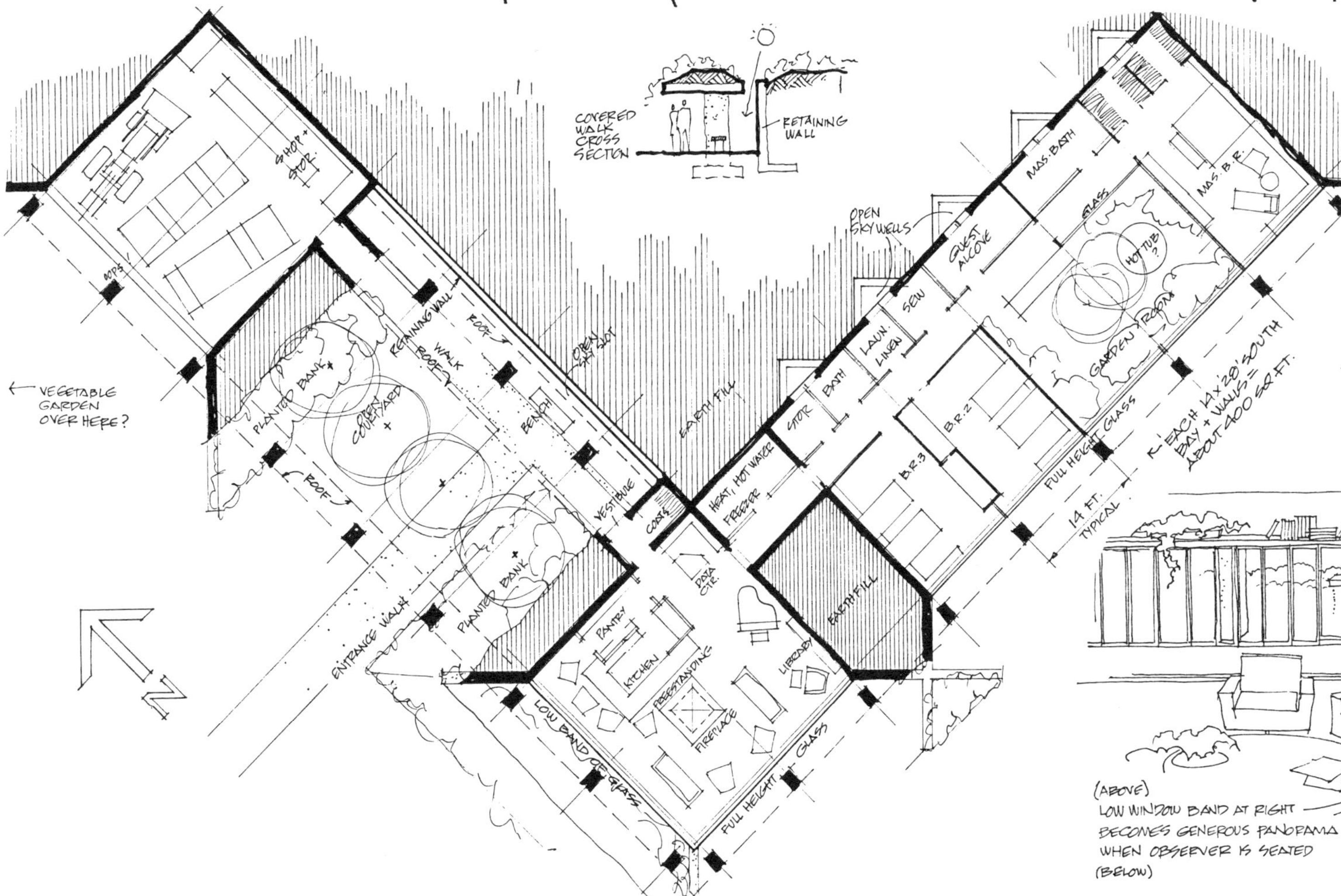

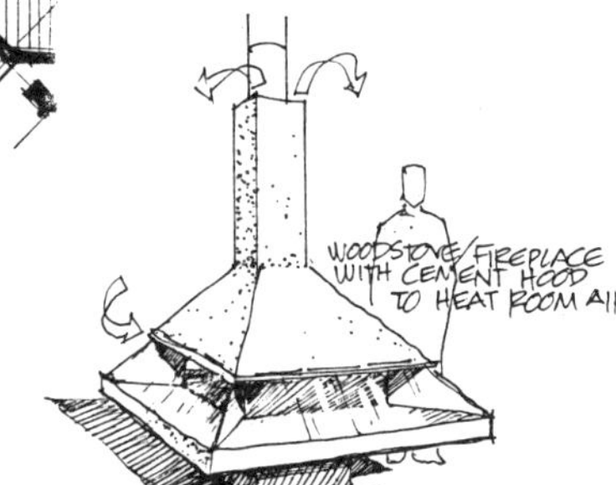

(ABOVE) LOW WINDOW BAND AT RIGHT BECOMES GENEROUS PANORAMA WHEN OBSERVER IS SEATED (BELOW)

odd-foot breezeway across the entry court. The site sloped down to the south, there was a lovely old avenue of deciduous trees through which the distant view was to be seen, and the plan was accepted with enthusiasm. The owners even liked the slit window at sit-down height along the west side of the living room.

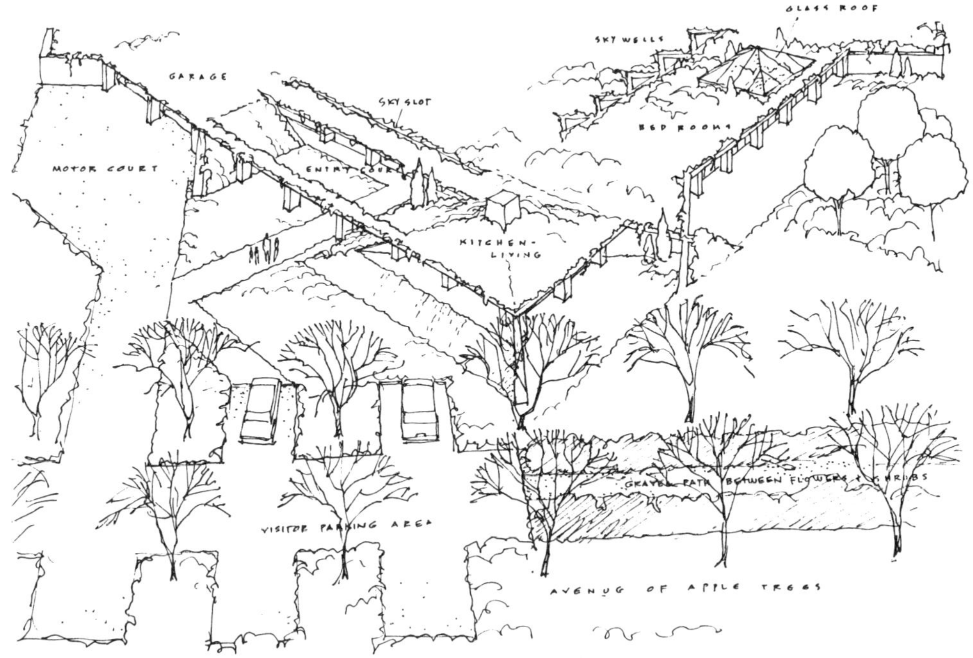

I know I told you, with great confidence, back on page 1, that most architects have relatively few jobs that go all the way through to completion but as I go through page after page of abandoned projects I'm beginning to wonder if I'm special.

Well, hang on, Mac, we'll be getting to some live ones in a page or two.

But first, we have this hillside solar earth shelter near Bradford, Pennsylvania, that never got any further. It was to have been part of a little bicycling community at which all the houses – ten of them, as I recall – would be strung along a horizontal path the main purpose of which was not to avoid having gentle slopes so much as to express, architecturally, the things about land husbandry that similar horizontals in contour plowing express.

Philadelphia was, for a while, to have been the center of the 1976 Bicentennial festivities. I did this proposal for the other side of the river, for poor old Camden, New Jersey.

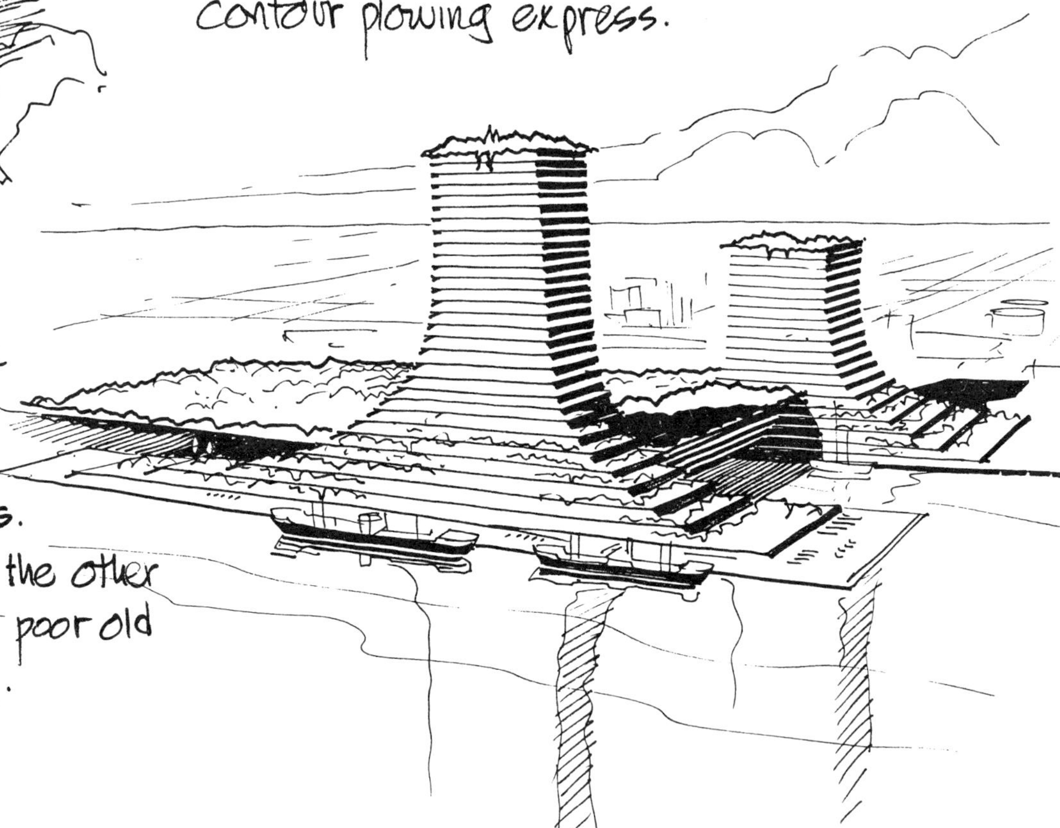

What *was* that thing, anyway?... a pair of office towers, or hotels, or what? You name it; that's what it was to have been.

Getting back down to earth again, we have, first, the sniffer duct, an idea that's worked successfully for me in several buildings. It simply pulls solar (or otherwise) heated air down from ceiling level, **before** it's wasted, and pushes it through under-slab air tubes, storing the warmth in the floor-mass before the air emerges, ready to go around again. I didn't use insulation below the slab in my latest building. My hope and belief is that the great mass of earth below will gradually warm up a bit and carry the building through the coldest of cloudy cold spells.

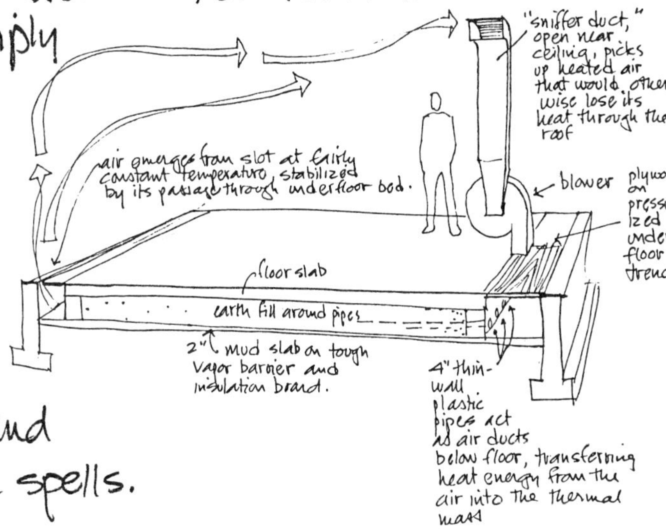

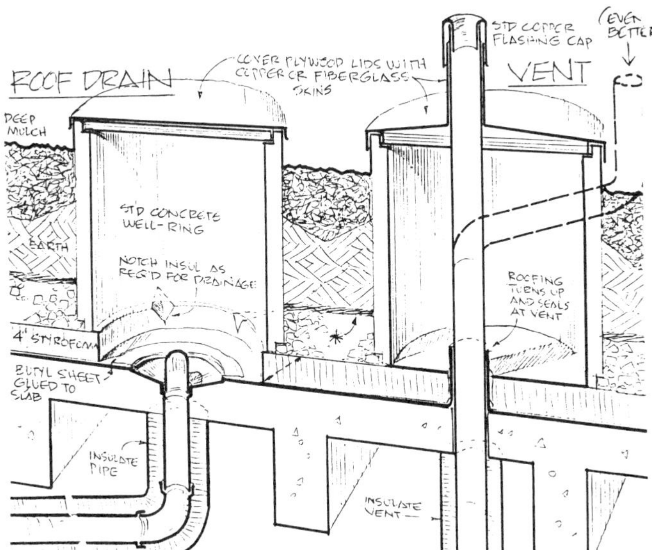

TWO ROOF-PENETRATIONS, SHOWING RECOMMENDED ACCESS TO EACH. THE LARGER THE INGS, THE BETTER.

It's nice to be able to get to roof penetrations without having to dig for them, don't you think?

This is my showpiece, or, I should say, this will be my showpiece.

Launched in 1971, "Locust Hill" at Raven Rocks, near Beallsville in eastern Ohio, is still under construction. Now in its 20th year, the great hillside house/office/workplace will be ready, this season, to receive its second floor. As you can imagine, there were twists and turns along the way.

In the early Seventies, the architect, still not persuaded that

solar heating would really work,
sited the house on an east-facing hillside, and several foundations were poured before energy expert Bruce Anderson and his company.

Total Environmental Action, got the owners to abandon the job. That they did so with good humor says a lot about the delightful people who've come to be among my closest friends.

Four people have been most directly involved in the Raven Rocks project but they've had a lot of help from the others in the 18 or 20 member group. The four: Chris Joyner, Don Hartley, Tim Starbuck, and Warren Stetzel. Over the years, Warren and I have traded hundreds of letters on every subject from the U.S.-

backed atrocities in Central America to the world environmental crisis, not to mention all the house details that have been proposed, discarded or refined, and worked into the project.

I will no doubt be given credit for designing Locust Hill but all the most important aspects of it have come from Raven Rocks. Hoping to make use of the very best thinking on what I call Gentle Architecture, they will not hesitate to travel halfway across the country in their VW bus to see a promising

experiment. Talk about open minds! They think nothing of changing plans in midstream (or however you say it) or tearing out a beautifully-completed piece of work if they think that in the long run (500 years or so) the building will serve as a better example if it incorporates more efficient features.

The Raven Rocks people have achieved a kind of self-sufficiency already, raising mouth-wateringly beautiful food in their lush gardens, repairing and

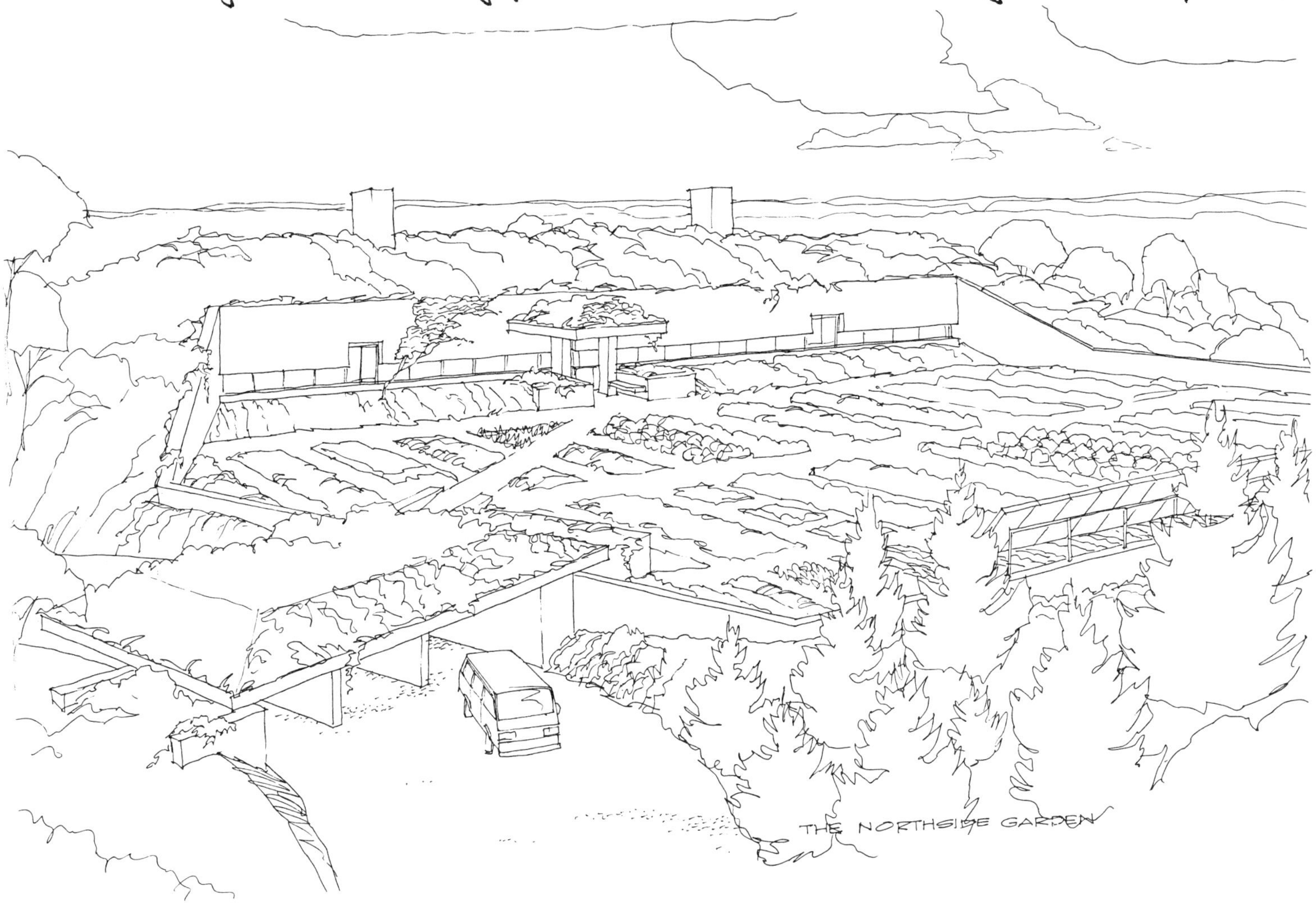

sometimes even fabricating parts for their equipment. They hope, someday, by the use of photovoltaics, to cut free of the electric lines that now power their temporary quarters on the site.

"Are they millionaires?," people have asked me. Far from it. They run a smallish concrete company, and they raise and sell Christmas trees (prize specimens, naturally, free of pesticides and even fertilizers). Money is not the main factor when determination is involved. Of course their determination is augmented by high intelligence, a world perspective, unfailing good humor, enjoyment of life, and sheer hard work. They think nothing, for instance, of staying up most of the night finishing concrete slabs if weather conditions, available time, and other factors indicate that that's the best thing to do.

UPPER FLOOR, NEAR THE SOUTH WALL

My daughter, Kappy, says I draw old-fashioned chairs. They don't look old-fashioned to me. Maybe 1950-ish, but is that old? I've been drawing Paul McCobb-ish chairs for 40 years now, not seeing, until she reminded me, that I've been standing still and the world's been moving by. So please don't be put off by my furnishings. Think instead about the often-diffused sunlight of eastern Ohio coming in through these specially-glazed, doubly-insulated windows and being trapped by them to warm the great masses of concrete nestled into the hillside. Think about the hundreds of acres of poetically beautiful ravines and forests out there, all land that the Raven Rocks people saved from the strip-miners with just minutes to spare.

May every architect have clients like the Raven Rocks group at least once in his or her career!

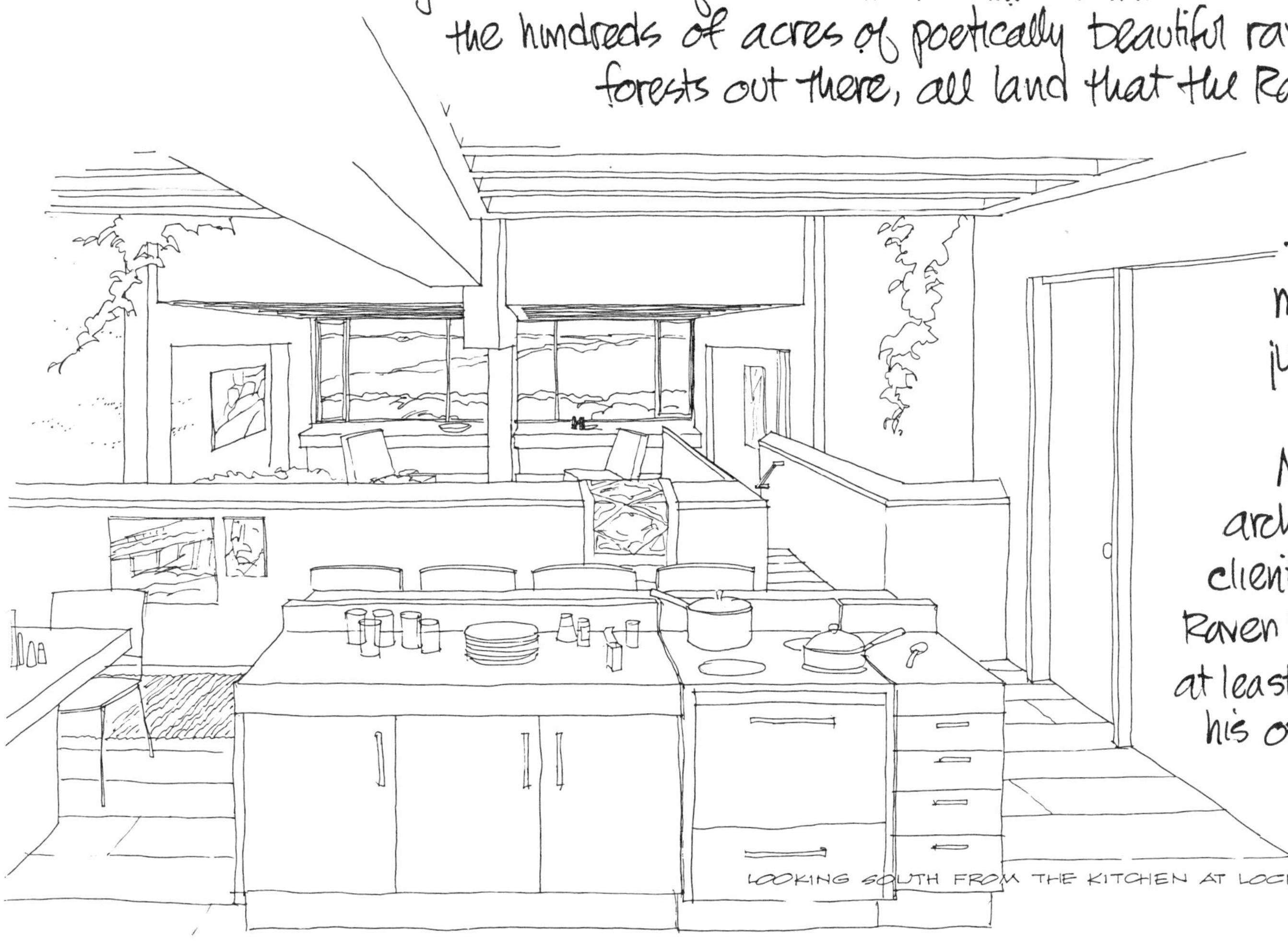

LOOKING SOUTH FROM THE KITCHEN AT LOCUST HILL

(Raven Rocks, cont'd)

Just down the lane from Locust Hill, Raven Rocks couple Mary and Rich Sidwell are building, too, at a considerably faster pace than that of their neighbors. Two years or so and they're putting the roof on already! Not having to duplicate a lot of the Locust Hill research helped, as did the construction experience gained over there.

Did you notice the carport tucked in underground below the Sidwells as they take a break for a moment and stand at the rail?

Clivus composting toilets have been used at all the Raven Rocks households for years, producing rich compost while the rest of us have been flushing our wastes into rivers and bays, or into fragile soils, as we do here on Cape Cod. We continue this insane practice even though the clean-up of nearby Boston Harbor is predicted to cost billions. I finally got a Clivus system approved and installed here in Brewster, but what a fight it took! The toughest adversary: the state Department of Environmental Quality Engineering!

But we all know that the wheels of progress turn slowly. My wheels of greenhouse-design progress for the Sidwells turned slowly enough. These drawings represent some → of the possibilities I proposed before being helped by the Sidwells in the direction of the design you see at left.

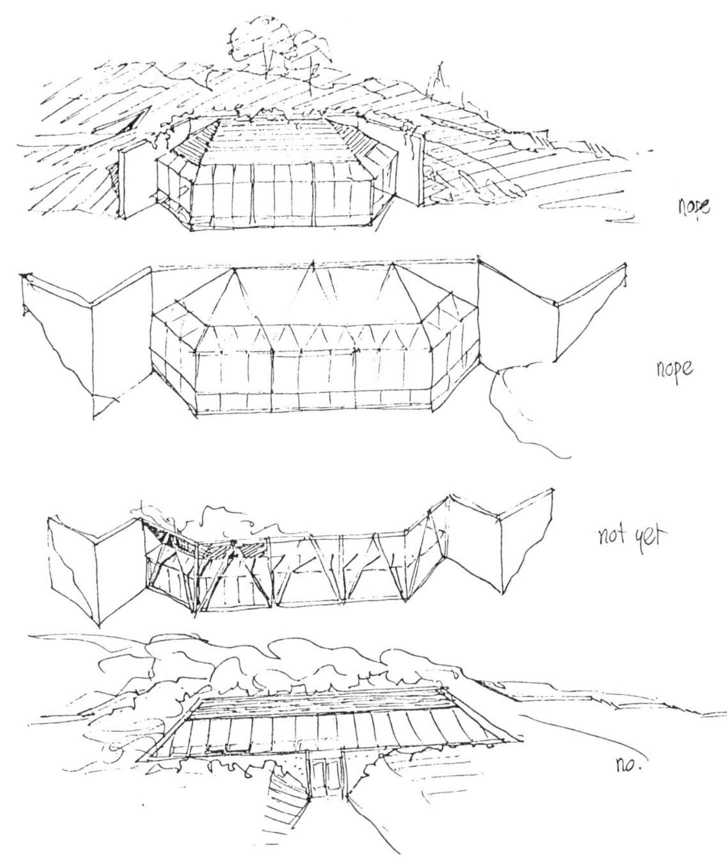

Once these Raven Rocks buildings get finished, and their landscaping has settled in for a couple of years, I'm going to spend a lot of money on color film.

I suspect, and hope, I won't be alone.

Here we have the south and north sides of the house I call the Hess house, probably because it was built for the Hess family.

The project went forward beautifully except for one thing: they decided not to put earth on the roof, so I can't show you a photo of it

with glorious rooftop plantings overspilling the sides.

But as a solar house it performed extremely well, and I picked up a bit more drawing experience.

Trained as an engineer at Georgia Tech, I never spent a day in its architectural school,* and my architectural apprenticeship was spent in the office of a man who did nothing but Colonial houses. It was good drafting experience but since the floor plans all looked like

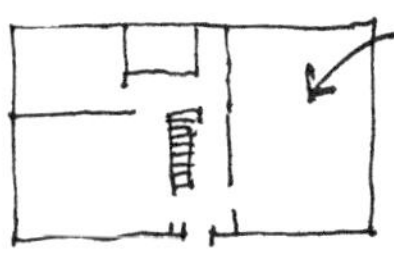

this, my own early designs never seemed to have any strength. Slowly, I learned to hang buildings on simple repetitive structural skeletons.

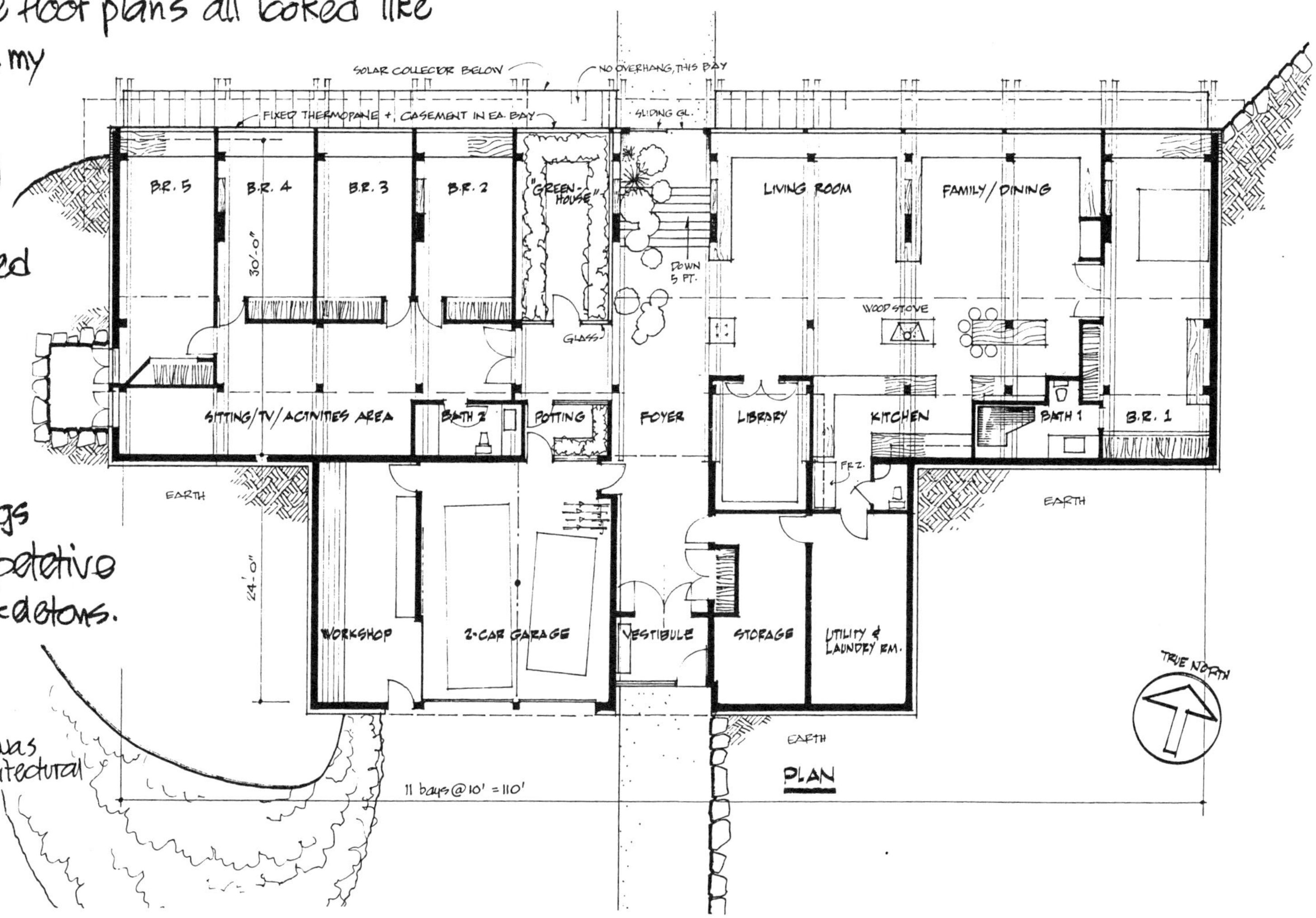

*except when I was dating the architectural librarian.

This skeleton was a bold one, designed to carry the earth load that never came along. In the perspective at right you can see that the fireplace has been moved back against the wall (on p. 91 it was free-standing). Other than that, the drawings appear pretty much to agree with one another.

Just don't try to blow under-slab air through a rock bed. Fortunately for me – and them – the Hesses never used that kind of heat storage system. Neither did any of my other clients in those (ca. 1979) days. Dust, cobwebs, and dead mice are just a few of the rock bed problems about which I've been told.

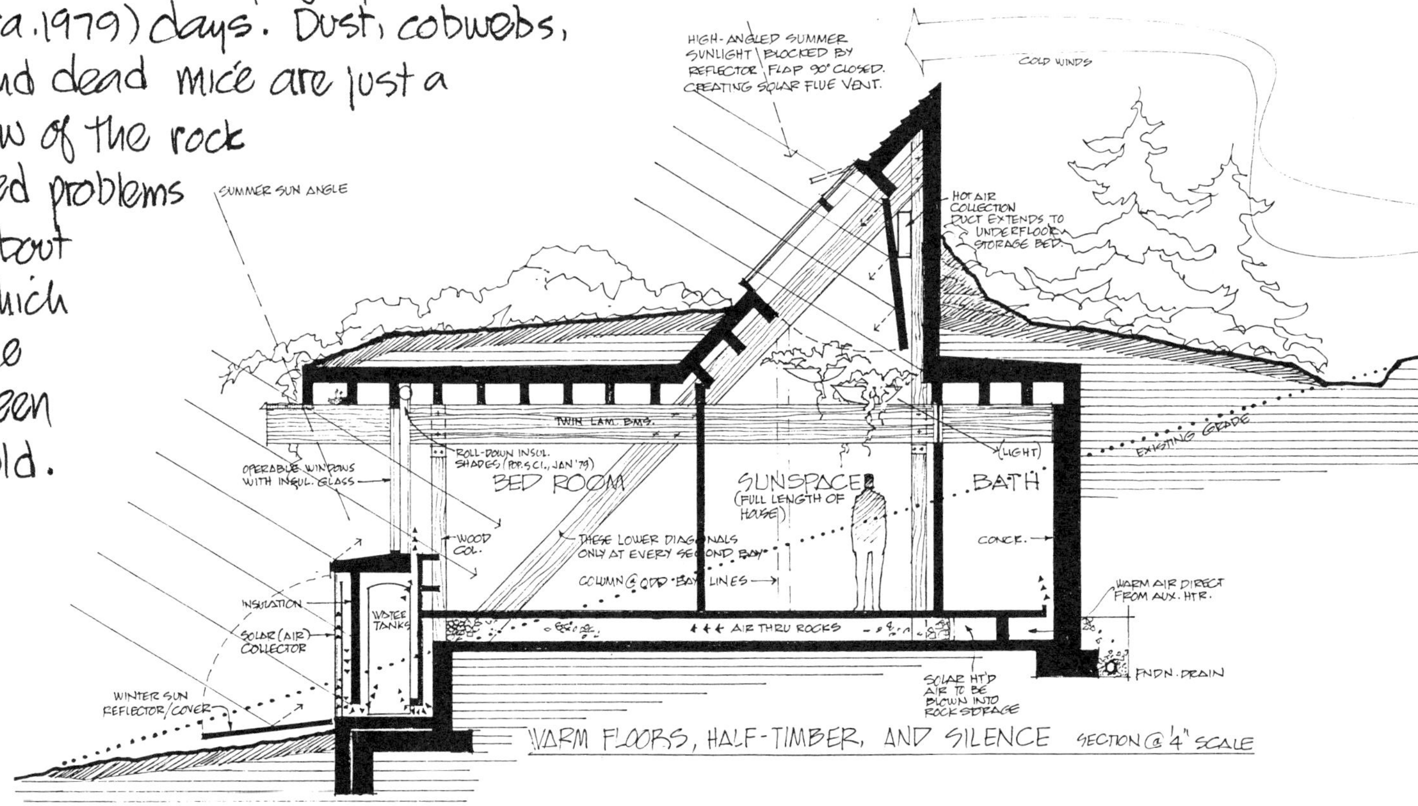

WARM FLOORS, HALF-TIMBER, AND SILENCE SECTION @ 1/4" SCALE

(Slightly more modern chairs, but only slightly.)

Cantilevered shelves have long been favored architectural devices with me but here they have a practical side as well), preventing unwitting head-bumpings on the sloping beams. When I do upper-room designs in the attics of more conventional houses, I like to let the roof and floor planes meet cleanly at a point. Cabinets or seats are always extended to head-clearance height in such cases, to avoid the bumping problem.

Location: somewhere in eastern Pennsylvania.

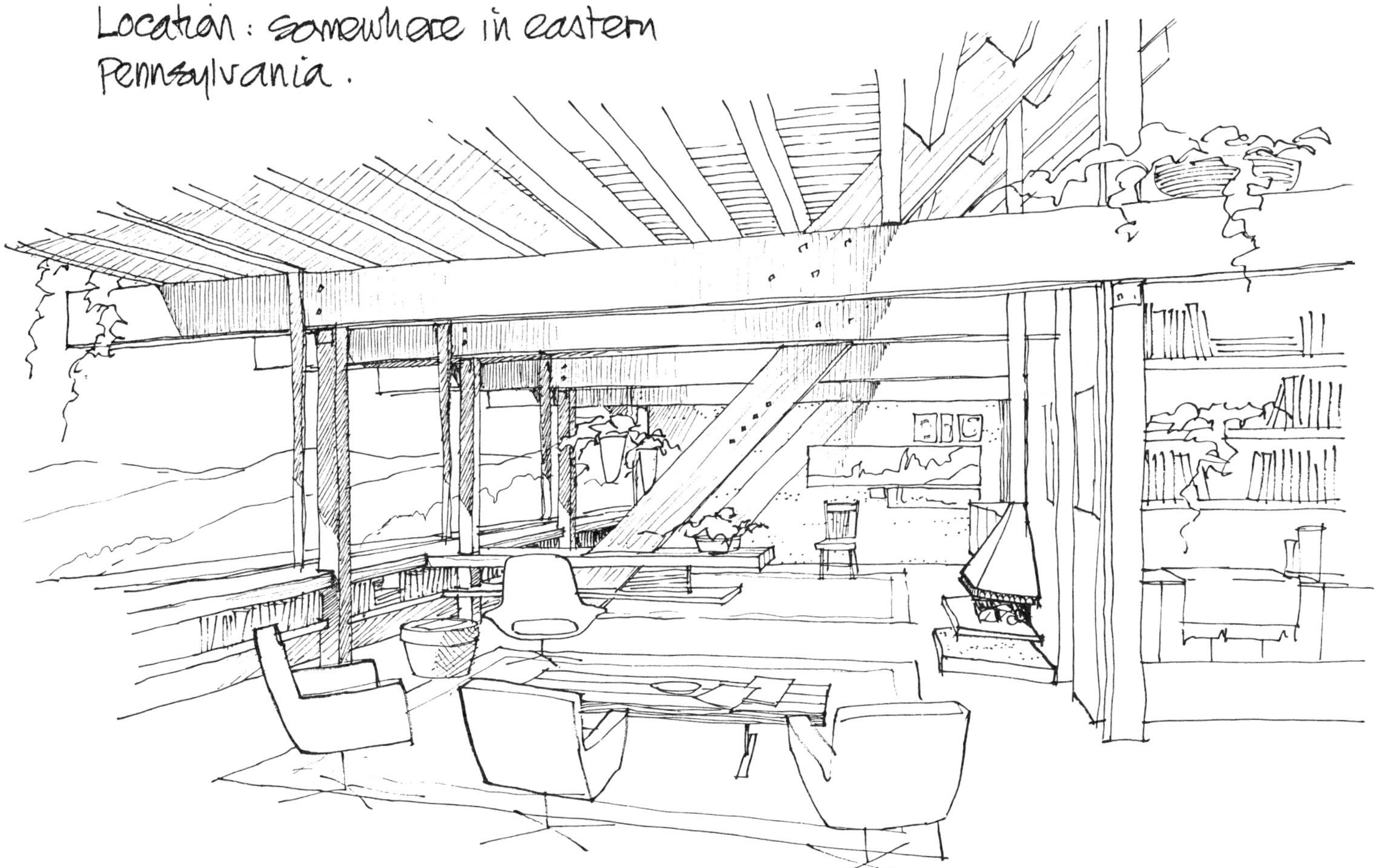

After 50 – 51 – years of living in New Jersey, and after almost 25 years of doing my share to contribute to its environment's downfall, I was able to move back into the best part of my past by moving to Cape Cod. My little self-published book, Underground Designs, was selling so well I was able to buy 23 acres of beautiful pine and oak woods, fronting on 3 ponds, for $78,000, cash! It didn't take long to find the best part of the tract: a high ridge overlooking the largest of the ponds. There wasn't another house visible in any direction. At night: not even a light.

The only trouble with that best part of the tract was that the ridge ran north-south (above), with the pond view to the east. The design solution: face south and tunnel through the ridge to a pond-view living room.

↓

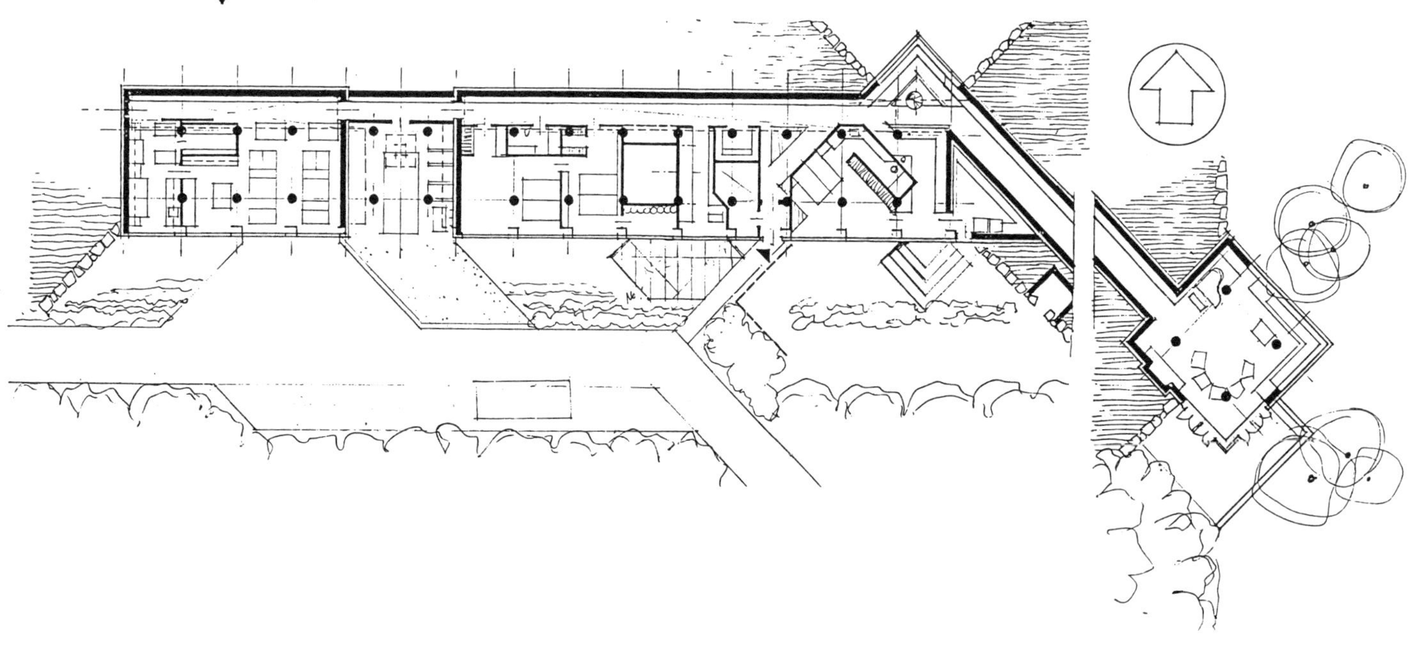

By facing every room south that way, the house got to be about a mile long. So, with visitors to Cape Cod already dropping in on our rental cottage in great numbers, it seemed wiser to put the visitors' quarters in a separate building and turn its back on us so they couldn't see what was going on all the time. That way, both our living room and the visitors' bed room (2) would have views, albeit angled ones, to the pond off to the east.

I hope to include, in the photo section, the perspective that illustrates this layout.

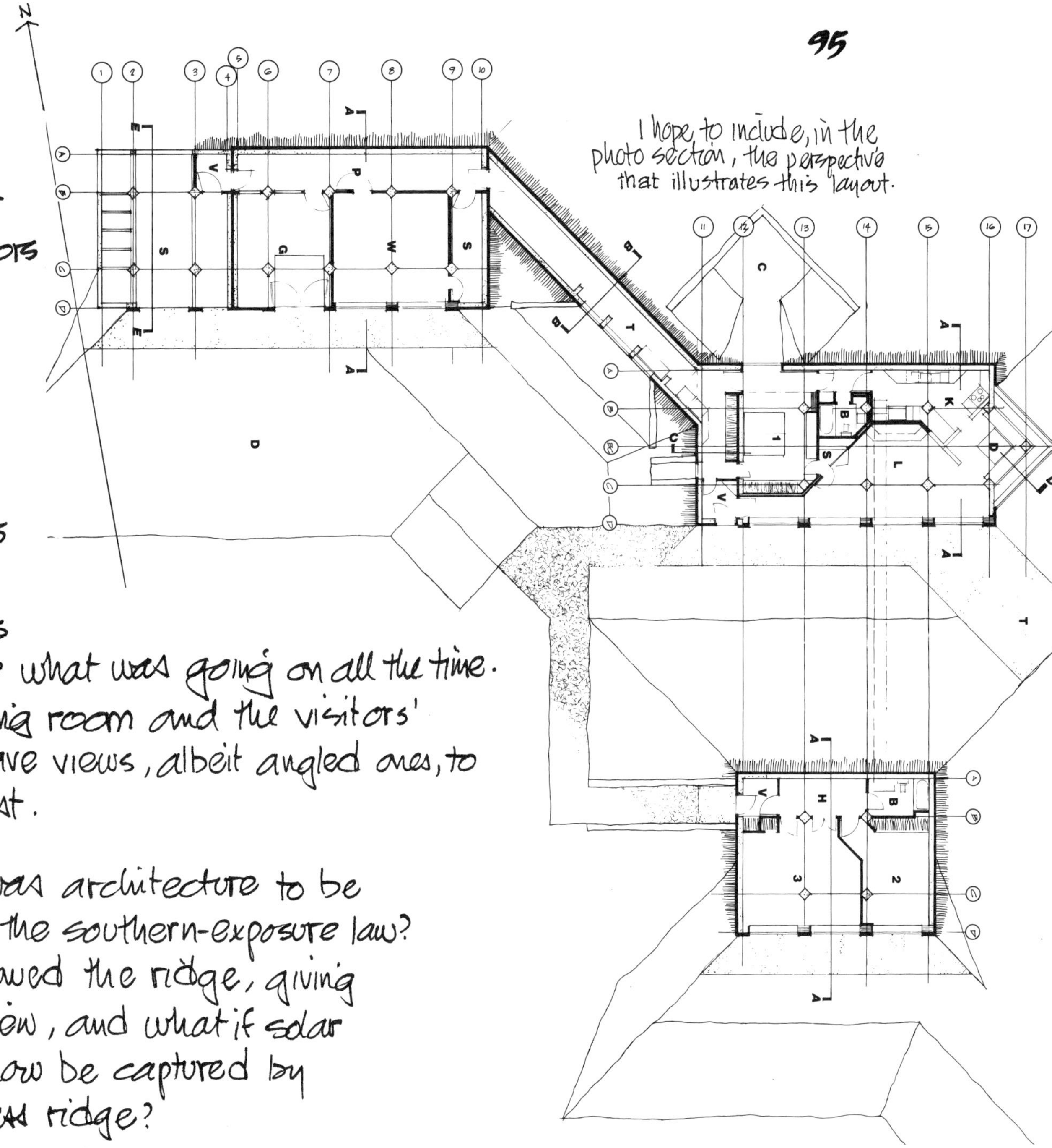

But wait a minute: was architecture to be forever shackled to the southern-exposure law? What if the house followed the ridge, giving every room a pond view, and what if solar warmth could somehow be captured by means of a long glass ridge?

So here it was, the ridge-line house ... living quarters to the left, guests and my office to the right. In the center, a glass-roofed – and unheated – foyer filled with plants hardy enough to survive winter nights on which temperatures could drop close to freezing.

Did it work? Truthfully, now. Yes, pretty well. The big double-glazed and window-quilted windows on the opposite side picked up a fair amount of early warmth; then the skylight did the job all through the middle of the day, and the high-up sniffer duct, right below the skylight, pulled the room-top warmth down and into the floor slab.

Cape Cod in the winter can have prolonged periods of raging winds off the ocean (9 miles from this site) with no sun visible for weeks. During such periods the little oil burner, combined with the effects of the great mass of insulated concrete and the earth shelter, was able to keep the house comfortable for a very low cost.

The main weakness of the house was its skylight design. After a few years the caulking in the wood members had begun to crack and admit water. And after those same few years I'd gotten sick of lifting the big winter-glazing panels (dotted line below) high up into the horizontal position and trying to bolt them into place while the weight of each panel was pressing on my head. A far

better system would have employed permanent, reflective-surfaced insulation-board panels – one in each bay – that could be rotated into and out of position either at the tug of a cord or by the action of a little motor.

But the principle, facing east but managing to capture plenty of solar heat, was sound, in spite of the bugs in the design.

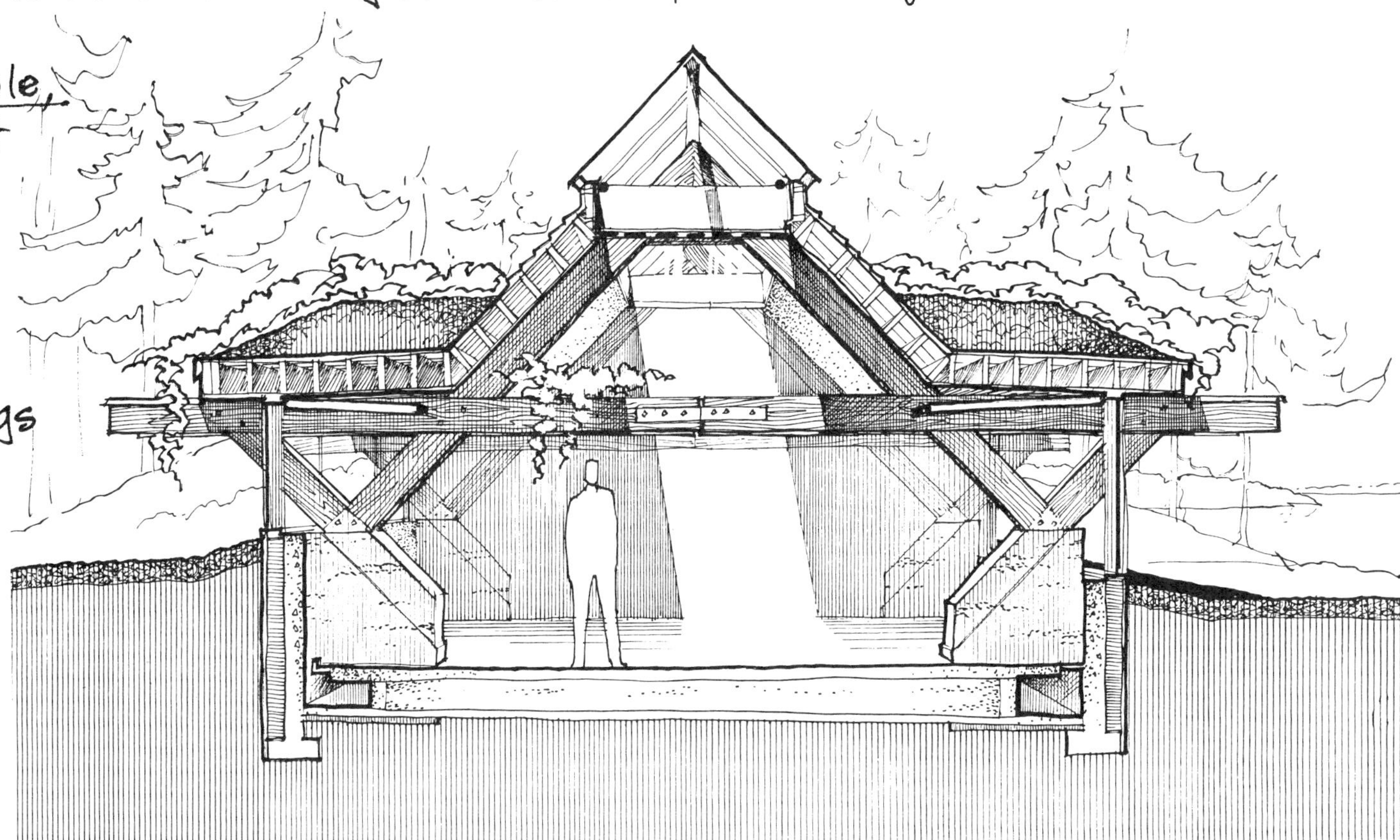

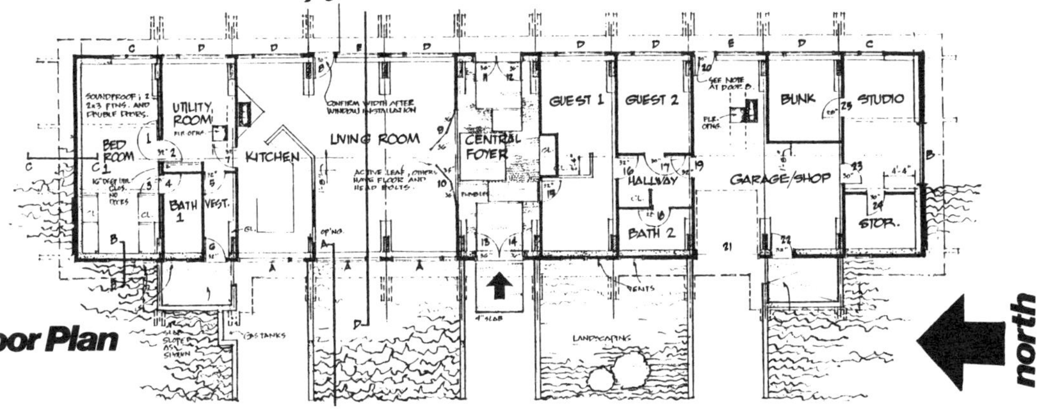

Floor Plan

The floor plan.

Each bay is 11'x 24', and there are 11 of them, bringing the total floor area to 7,692.4 square feet.

No, that's not right; let me try that again: 11 x 11 x 24 = 2904 square feet. It cost less than $50/sq. foot. Today, 1990, about ten years later, the cost would be more than double.

It was a nice house in a beautiful location, but troubles, other than architectural ones – or even financial ones – put the house in the hands of others, and it has been changed considerabley as a result, its skylight removed and shingled over, its sniffer ducts gone, its rooms rearranged, and its garden circle bulldozed away.

I wouldn't be surprised to hear from the new owners at any time. "What are all these trenches in the floor?" they may ask, or, "why are there terra cotta flues coming up out of the plants in the foyer?" I'll try to behave myself, if that occurs, and not make up any fantastic lies about nuclear reactor vents or mink-breeding tunnels.

After all, the house is only a mile away and I want to be a good neigh-

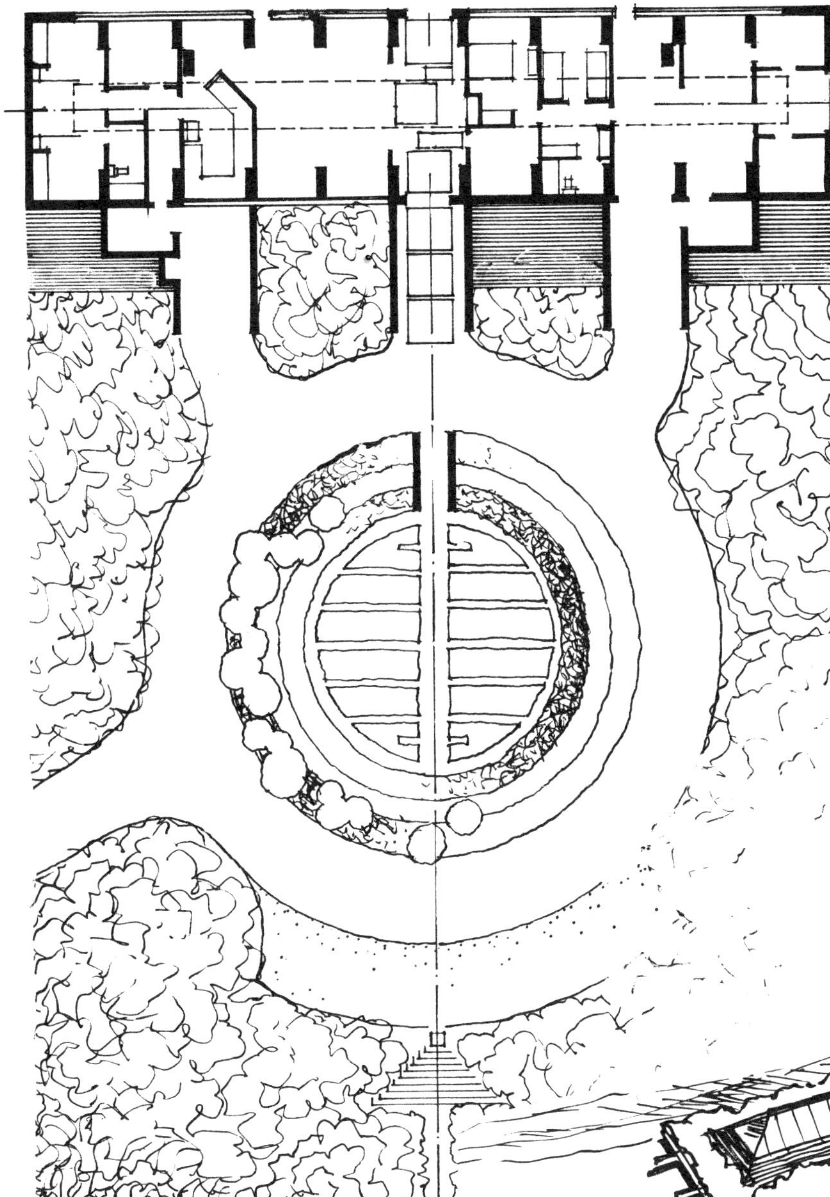

bor.

The garden circle I mentioned was, on completion in 1980, an afterthought. "Do you want this pile of fill taken away?", the contractor asked.

Hmm...

"No," I said, "please spread it to form a circular mound. Then I can have a vegetable garden that's hidden from view!"

And that was that.

Surprisingly.

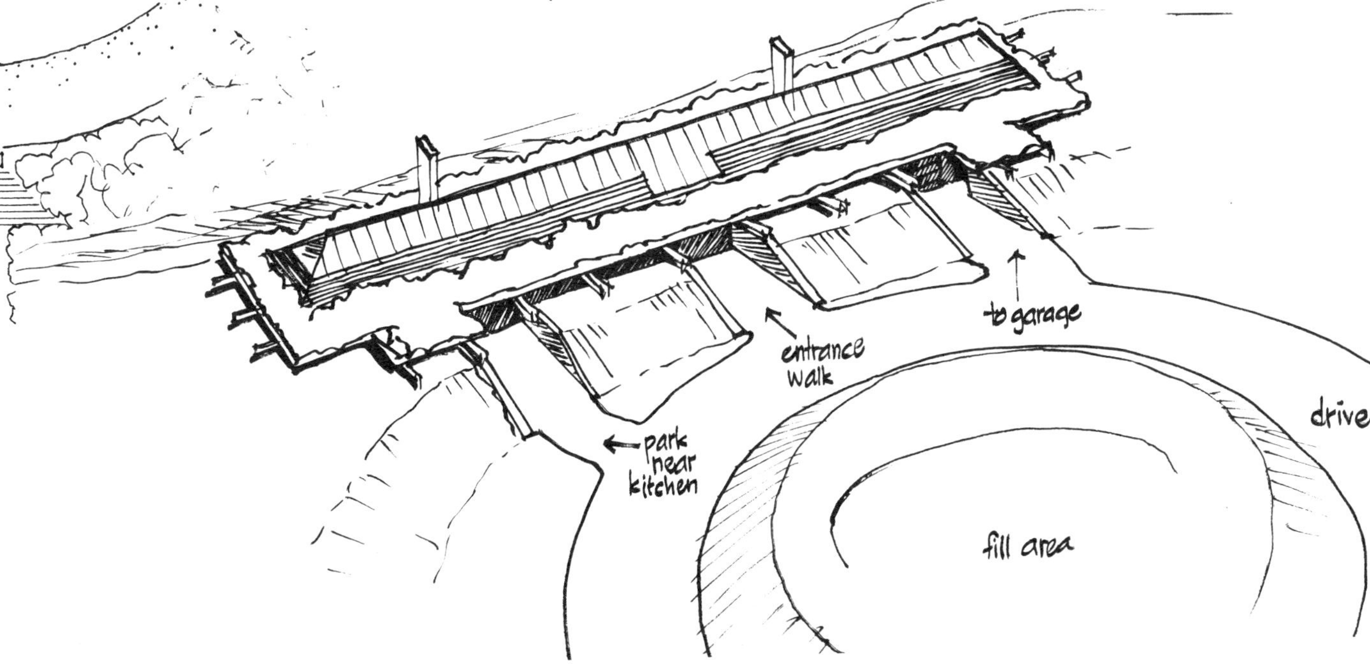

it not only proved to be a handy spot for growing vegetables, it proved to have a much more stable temperature, probably because of the encircling earth mass, than any garden I'd worked before. In the fall, when local gardeners were having frost troubles, I wasn't. It was often weeks later when frost would have the nerve to enter the circle.

Our little furry friends in the forest never did have the nerve. While others complained of the damage done their gardens by rabbits, squirrels, muskrats, and deer, we couldn't complain of any. All I have to explain it is an amateur's theory: animals, which surely checked out the living salad in the big ring, felt they'd be unable to escape in a hurry and, as a result, were never quite tempted to sample the fare.

With amazing foresight, I devised little inverted pans to protect the vulnerable top surfaces of the projecting twin 3x12 pine collar ties. Each pan had an upturned flashing that went behind the fascia (at "x"). The wood was like new after 5 years.

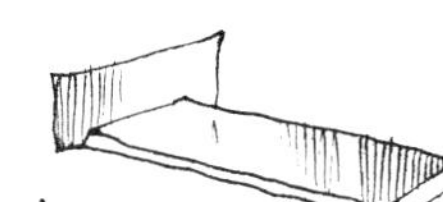

But I never anticipated the air leakage all around the 3x12s where they passed out through the cement-stucco'd building wall (y). During the first howling nor'easter in the winter of '80-'81, the drapes almost blew down from the force of the wind coming in all the cracks around the 3x12s. Caulking helped, later on, but as the members continued to shrink, and twist, new openings appeared.

There is no 100% solution, particularly if the wood checks, that is, splits harmlessly along its length. But if pieces of sheet metal are inserted into 1/4" deep kerfs (saw cuts), most of the air leakage due to shrinkage and twisting can be eliminated. The sheet metal occurs just at the outer face of the wall sheathing.

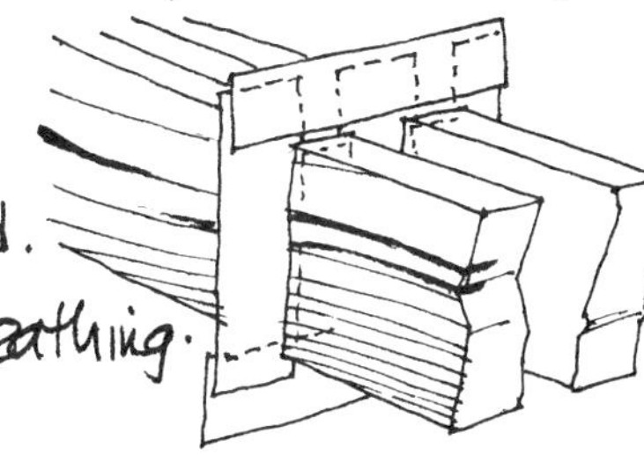

And now for a final look as we lift off and fly away, this view being from the south.

Don't miss the photo pages!

I feel just a little ashamed of this drawing, now that I see it again. It was my first response to a commission to build a solar, earth-covered addition to a house in Greece.

Imagine: greece! what a lot of impressions that were generated!

I'm embarrassed because, having been given the design of the existing building I failed to bow to its precedent and let a blocky rectangular theme run through all the rest.

Yes, there's a nice solar/hillside statement here, but if I missed the boat on the basics what good is all the rest? How could I do that to the Parthenon people?

Fortunately, the question was moot. Nothing ever came of the first exchange of letters.

then why are you wasting our time with all this nonsense?

Just in case the lesson is of use to you.

The Nature Museum; Charlotte, North Carolina. 1976

Now *there* was an opportunity. Underground parking. Underground building. Solar energy. Natural plantings. Rainwater percolation. A nature-oriented client. Exhibits. Education. A downtown location. A willingness to experiment...

Whatever happened to that job, anyway?
Beats me. I suppose it was awarded to someone else, designed, built, and admired by all who saw it. I hope so. The people I met at the little building that formerly housed The Nature Museum deserved a good one. Their values were, to my mind, right on the money.

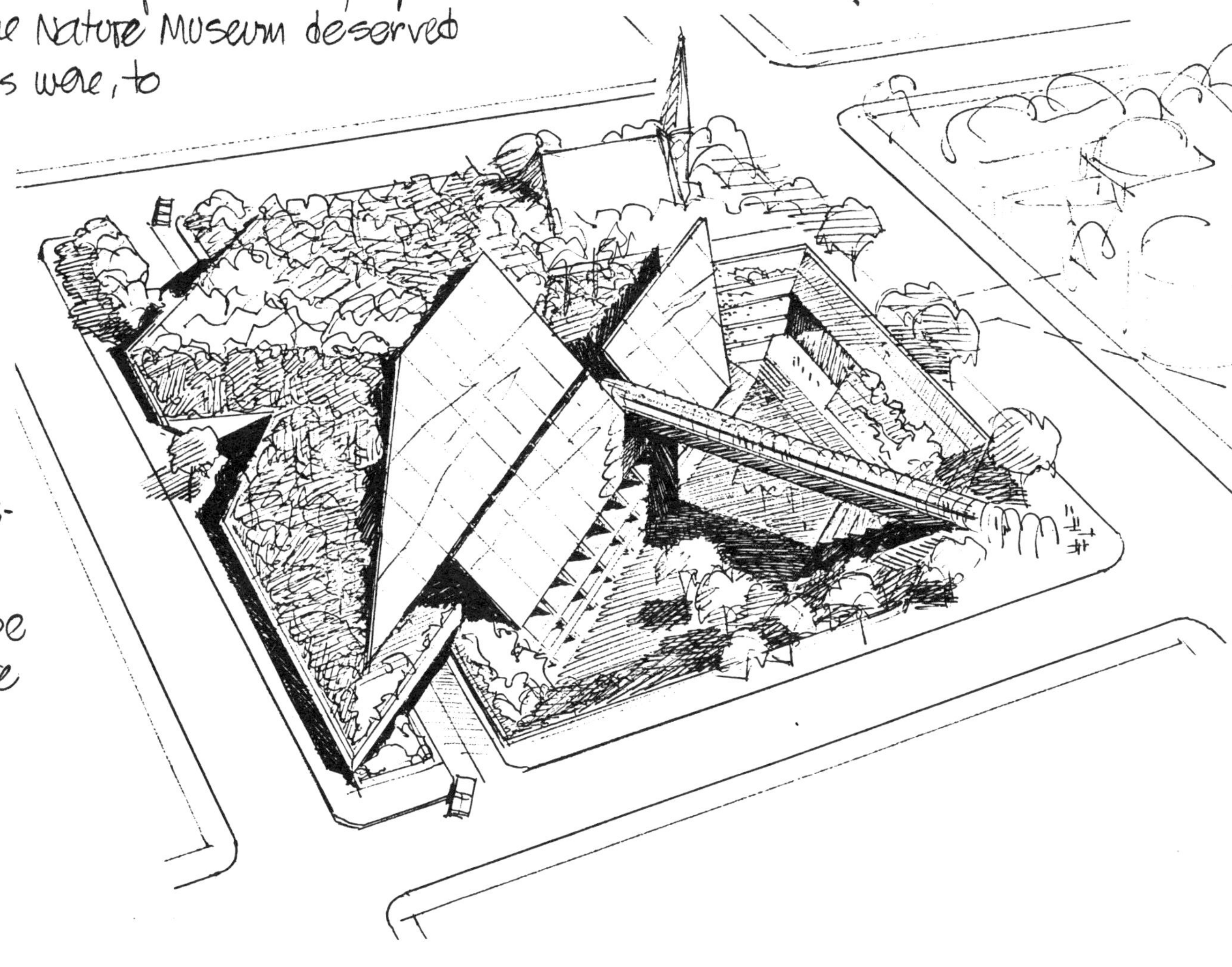

Maybe the existing church, there on the other corner of the block, would have been lent some ethical reinforcement by the museum. And, of course, maybe the museum would have been lent some by the church.

How's this for a simple floor plan? It's a cousin of the one I'd been playing with for years, finally to appear as the Popular Science house (pp. 184,5) of April, 1989. A later version of this plan, before this project, like so many others, was abandoned, had additional rooms in what is shown here as an open-to-the-weather carport space. The extra space would undoubtedly have been useful but the nice outdoor separation – putting the kids, or the guests, off by themselves, not to mention the appeal of having a deep, inviting recess in the center of the facade – would, in my view, have been the better choice.

This went up to a vegetable garden on the gentle slope north of the house.

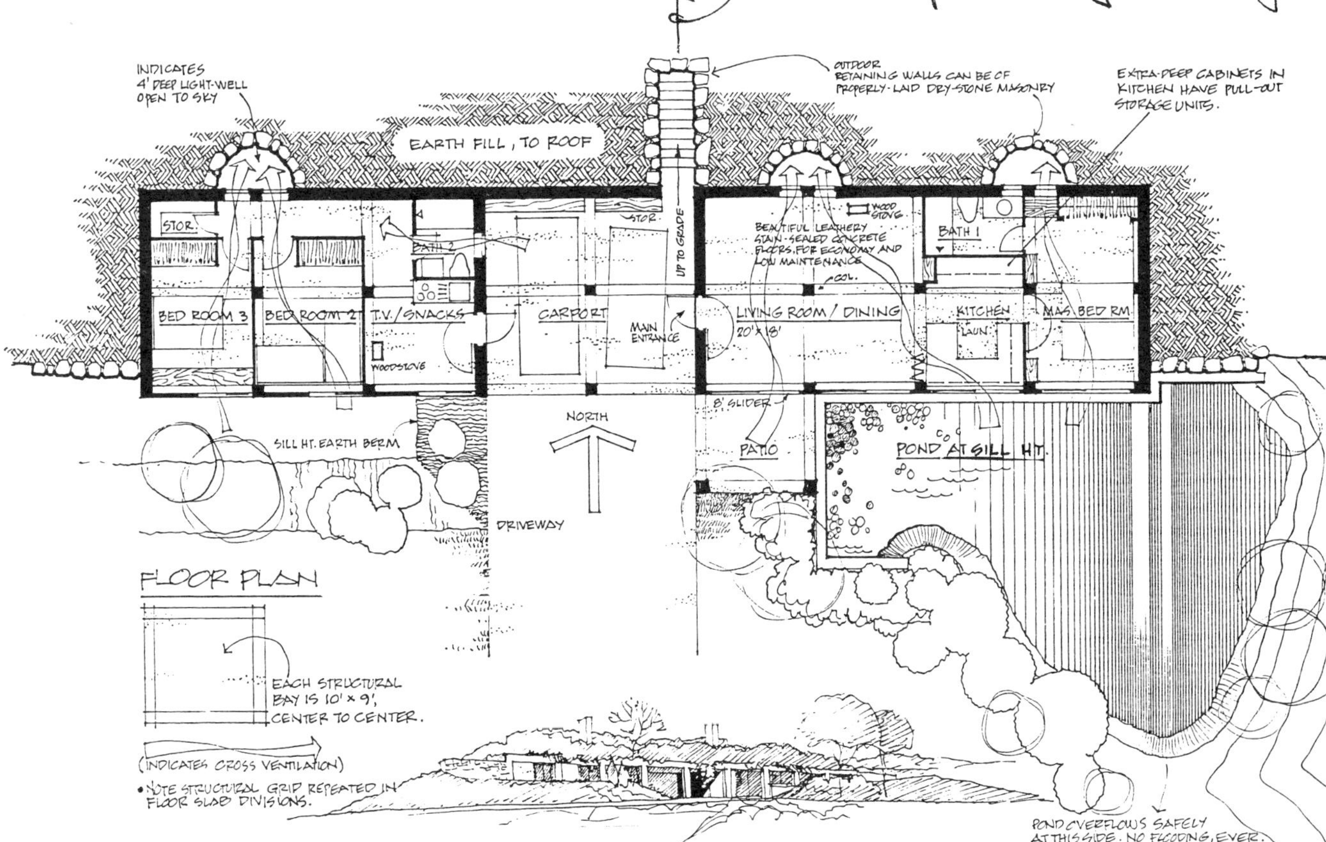

Sunlight, burning wood, and sniffer ducts would have handled the heating.

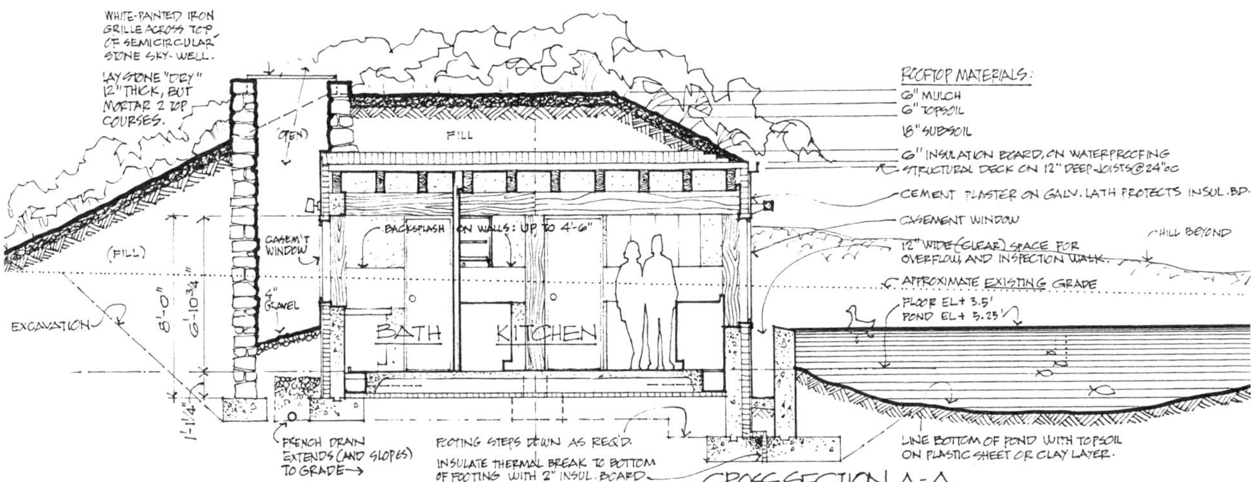

CROSS SECTION A-A

You wouldn't want to get stuck in that narrow overflow slot between the pond and the house! Better to make it, say, 24" wide. The timber trellis, shown below, was to have covered only the little patio but if the posts had tiptoed right across the pond, shading all the east-end rooms, it might have been more effective – and dramatic.

A problem to consider: if the last (northernmost) purlin (sub-beam), shown resting on the concrete wall, shrank 1/2" in height, and the other purlins (and the main beam supporting them) shrank a total of 1", would water forever be trapped in/on an underground roof puddle? Would it matter?

Better slope the roof a little.

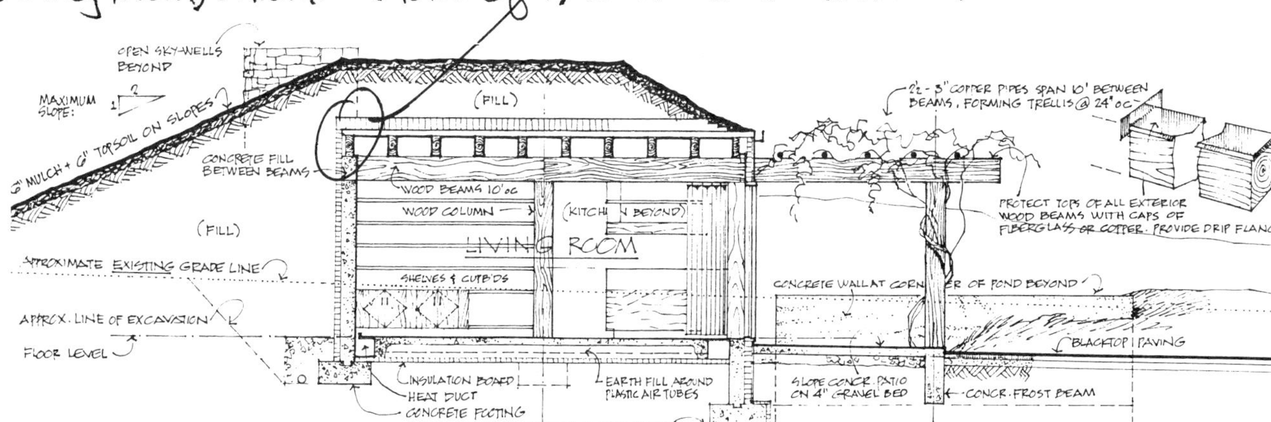

CROSS SECTION B-B

A RESTAURANT THAT LOOKS RIGHT DOWN THE LAKE, EXPRESSING, IN ITS FORM, THE IDEAS OF "VIEW," "SOLAR ENERGY," AND "LAND HUSBANDRY."

I SENT THOSE NEWLYWEDS UP ONTO THE SLOPE TO SHOW YOU HOW BIG THE DINING ROOM WINDOW-BAND IS.

If you can't bury them, berm them, isn't that the rule? When I was asked to design a restaurant and a bank building at an office "park" surrounding a quarter-mile long lake, I knew, from having tried to sell the idea there before, that earth-covered roofs would not be accepted. All I could do was point in the direction of earthiness, with some plantings on one roof and landscaped slopes up to the other.

Writing about roofs reminds me that I've said very little about the waterproofing of earth-covered roofs, so, if I don't mention it in the next few pages, please remind me.

NOT JUST ANOTHER BOX-BUILDING ON A BUSY STREET CORNER BUT AN EARTH-OBJECT; PART OF THE LAKESIDE WOODS, INVITING ITS CUSTOMERS TO A BIT OF SOLIDITY AND PEACE.

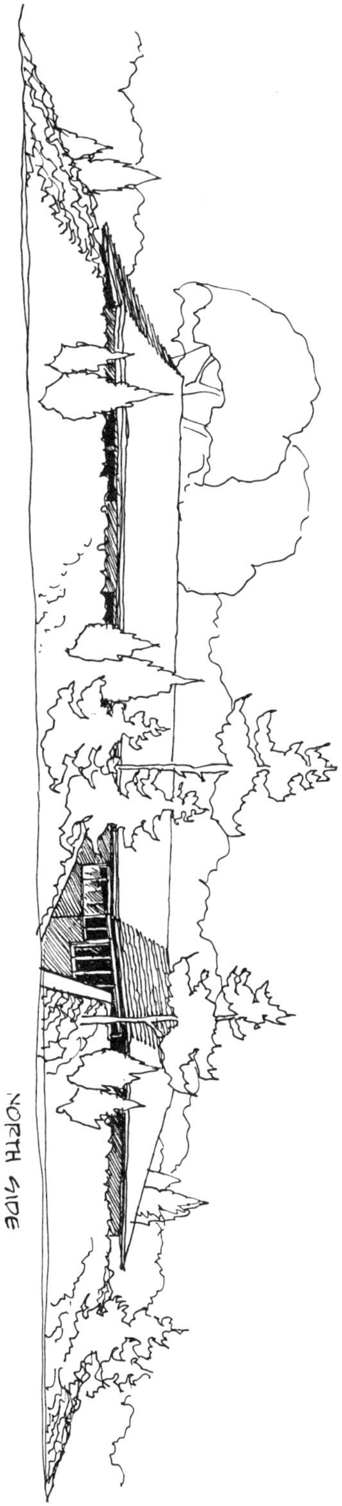

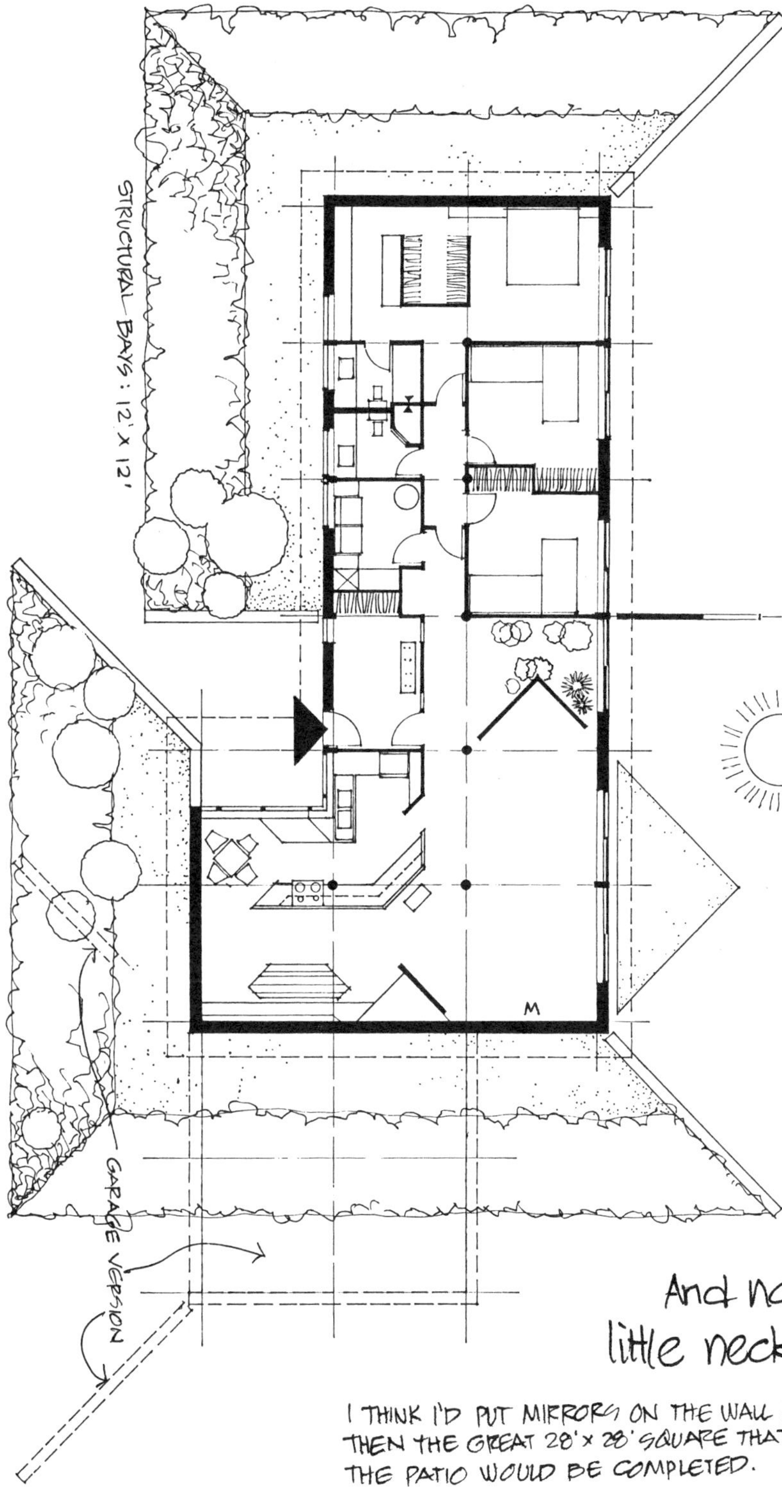

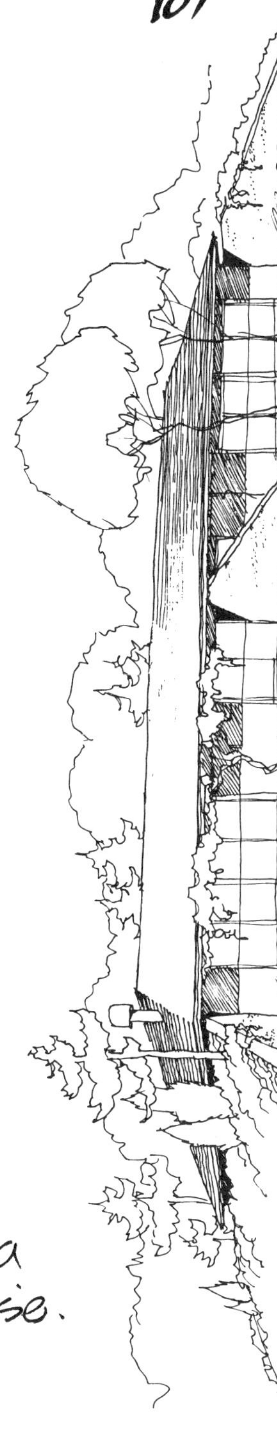

And now for a little neck exercise.

I THINK I'D PUT MIRRORS ON THE WALL MARKED "M". THEN THE GREAT 28' x 28' SQUARE THAT INCLUDES THE PATIO WOULD BE COMPLETED.

Cape Cod is a lovely peninsula of sea and sun, and forests of pine that are being stripped away at a frightening rate to make room for thousands of phony cape cod cottages, and acres of shopping centers that look like phony houses.

I've had two chances to build a home and office for my wife and myself, and I'm more determined this time, perhaps, than I was the first time, to resist the new cape cod tradition.

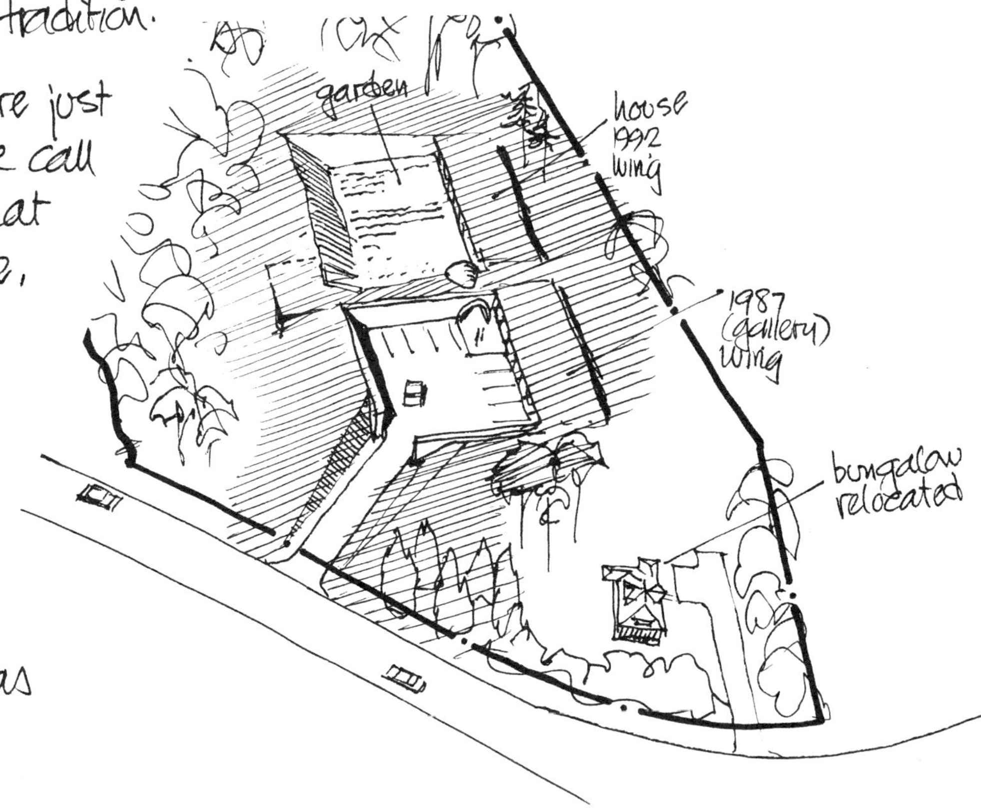

My wife, Karen, and I are just now completing what we call the gallery wing of that second Cape Cod house, and these drawings illustrate its evolution.

The new gallery is barely recognizable in this 1985 version but its basic 60'x20' underground shape has never changed.

I've long wanted to have a low (5 foot) ceiling area along a north wall in order to allow the rooftop earth to slope down low against the winter winds, and to do away with the high, blank facade usually seen above the small windows on the north sides of buildings, but no, the code won't allow low-head spaces even though they are taken for granted in our cars.

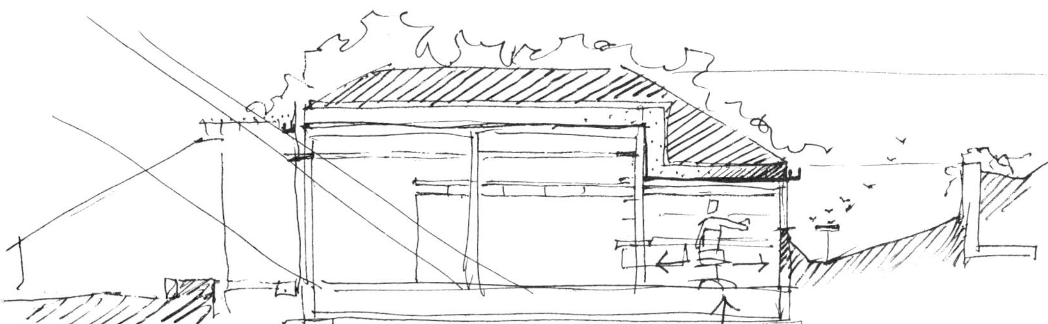

(The chair would roll into the low area.)

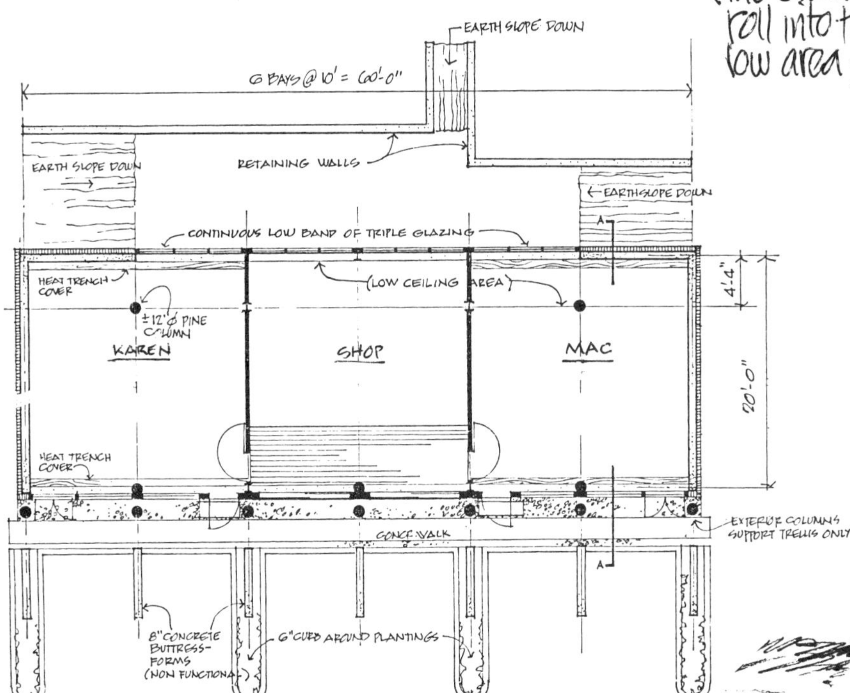

FLOOR PLAN
1986

THE UNDERGROUND GALLERY

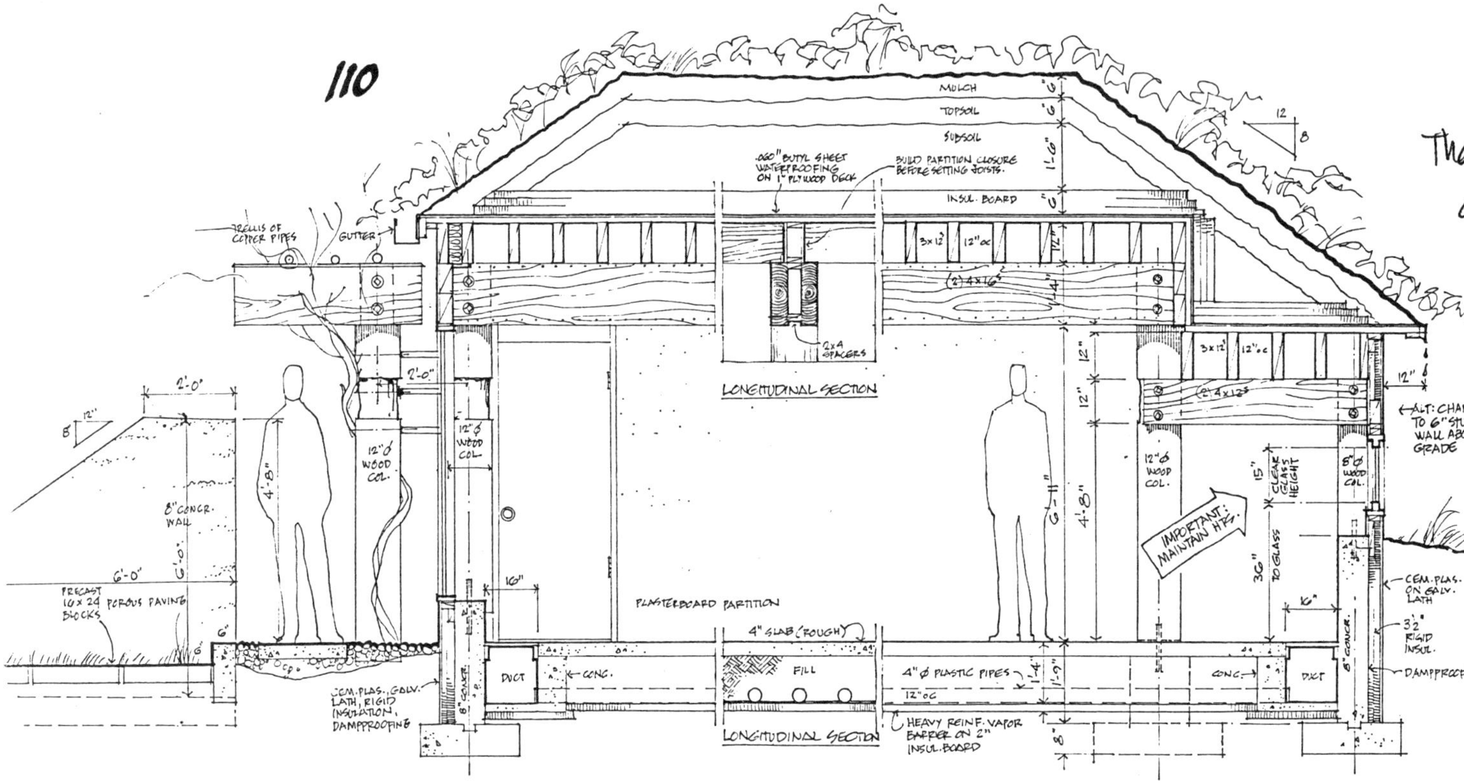

That carved recess around each column near eye-height would have been nice, and we still may do it. But on this page, design evolution is the subject.

CROSS SECTION, 1986

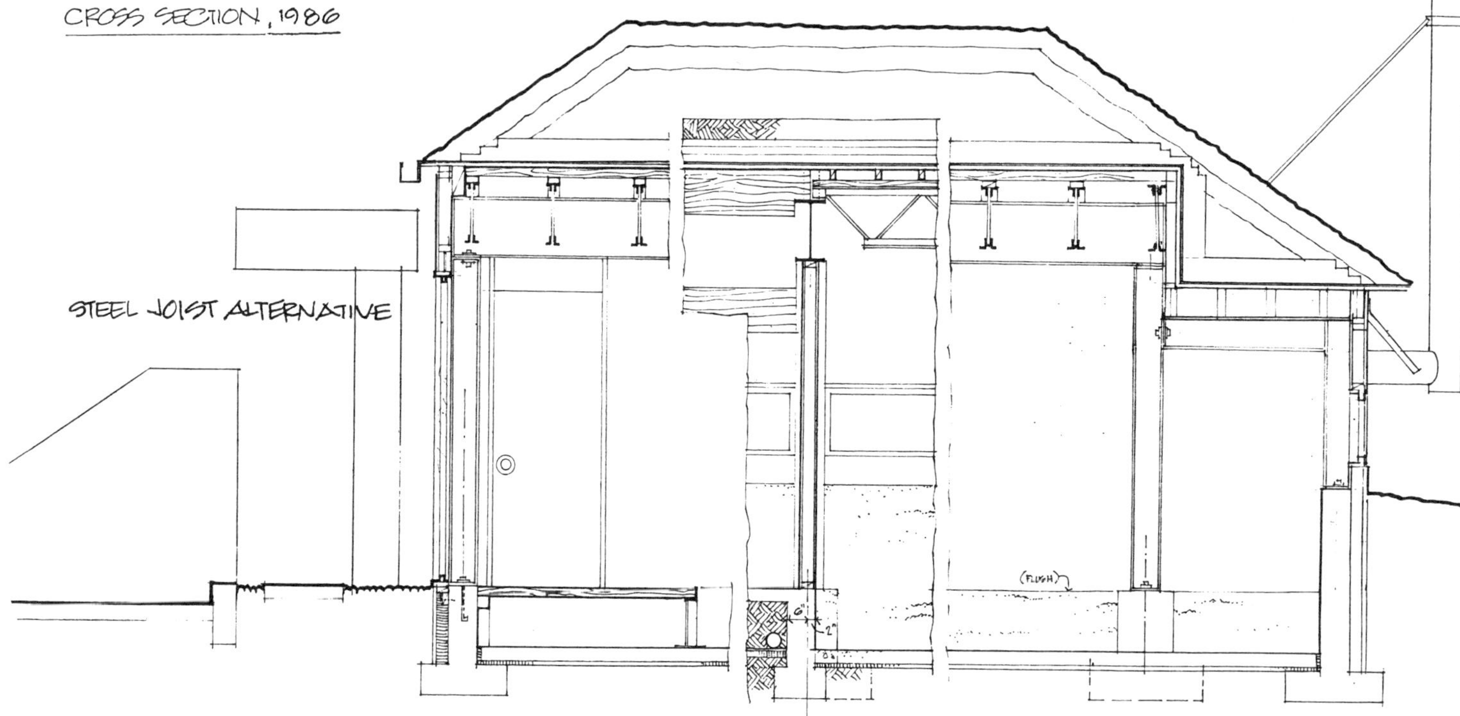

STEEL JOIST ALTERNATIVE

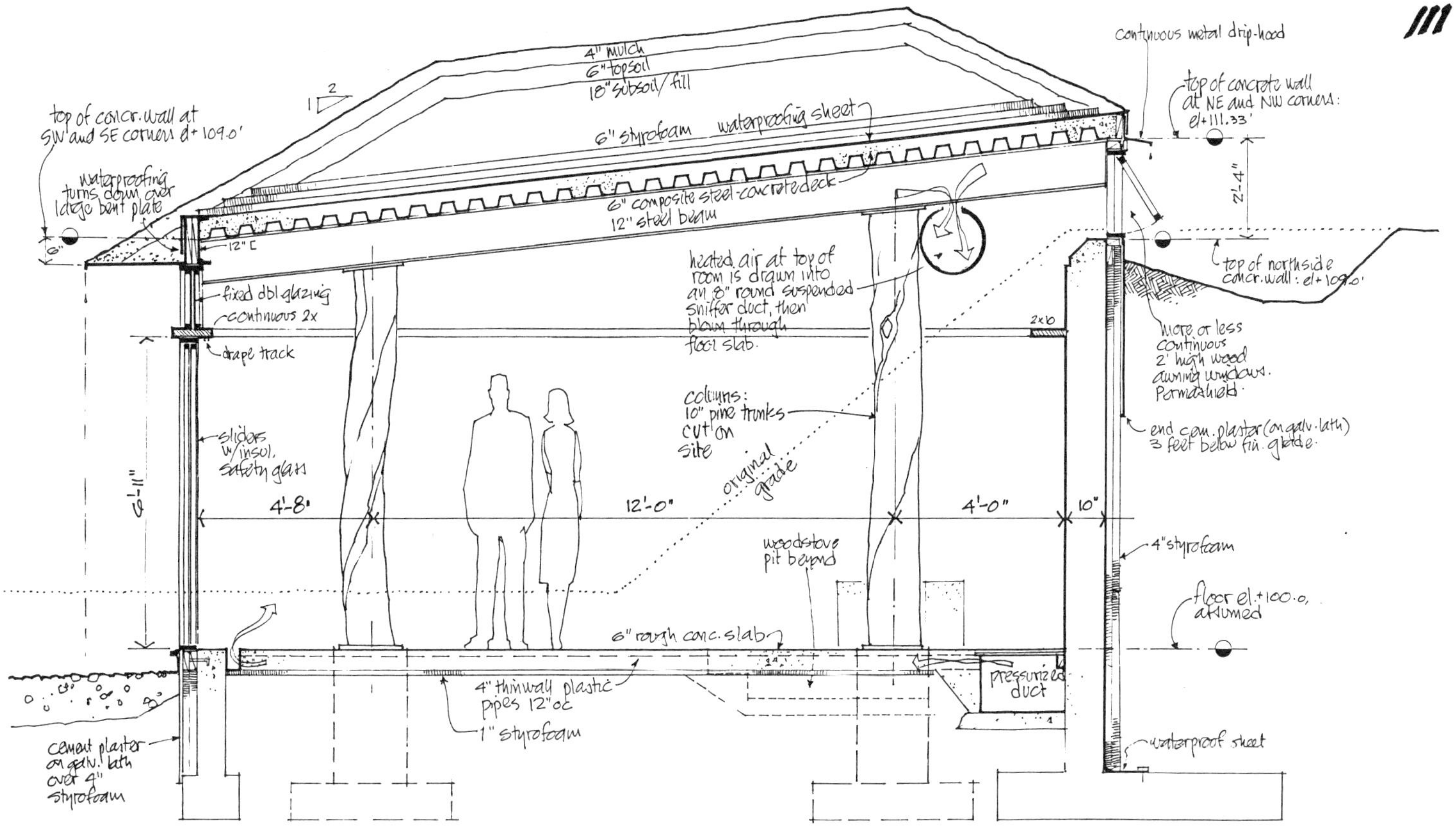

These are the as-built section, eaves detail, and floor plan. Opening date: as soon as a lot of @!# legal hassles have been resolved.

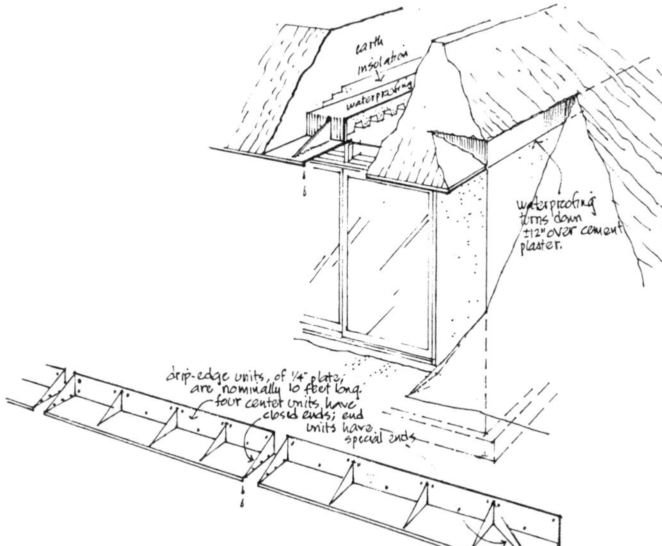

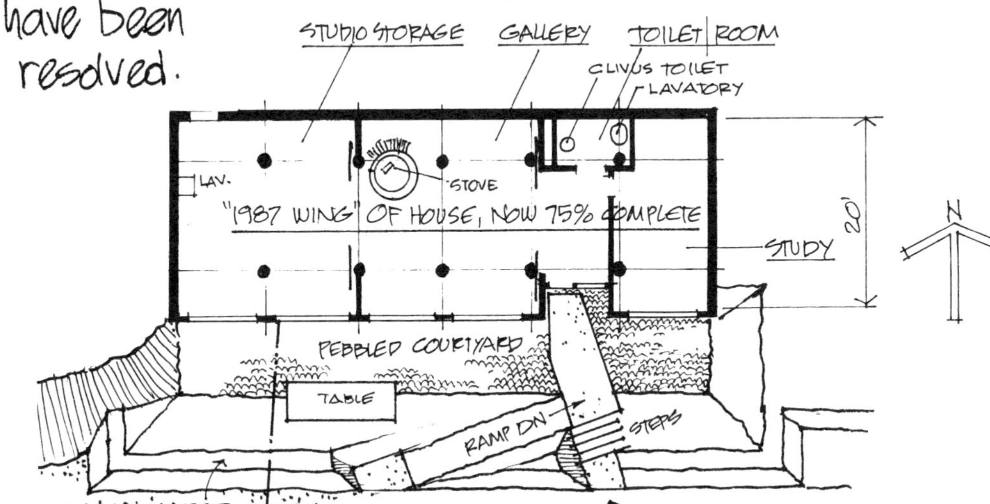

THE UNDERGROUND GALLERY MW 3.7.88

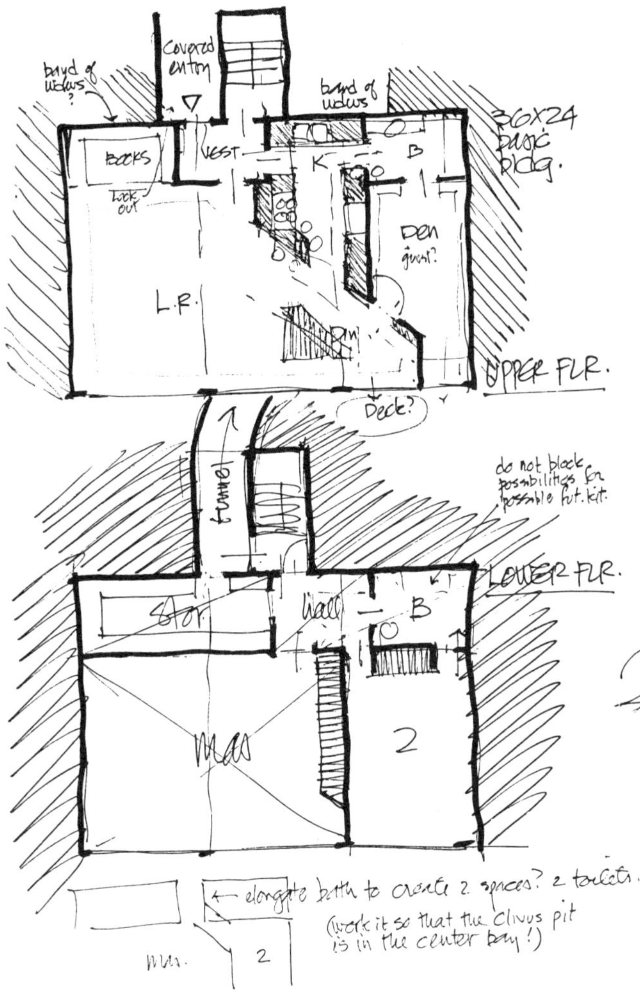

The 1992 "living" wing of the gallery is still under discussion. How many bed rooms?

Can we direct visitors to motels? How many stories? The idea of a tunnel, down at gallery level, linking the wings, has great appeal to us but the cost does not, so we keep fiddling with the plan. The house has to be solar heated, earth covered, and thermally efficient.

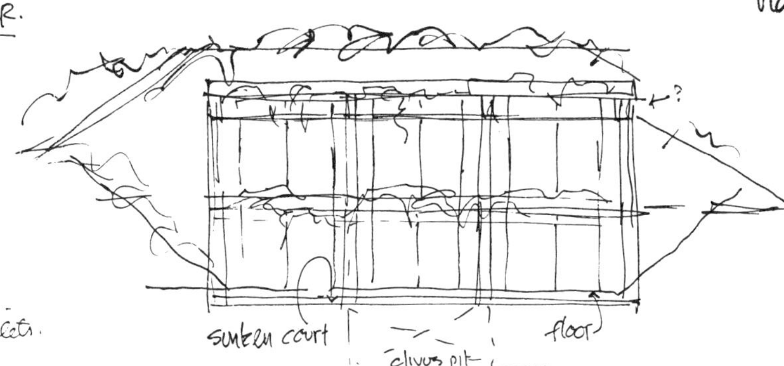

Other than that, it all depends upon the size of the mortgage.

What appears at right to be a three-story version of the house is actually four; there has to be an underfloor space for the sewage composter. Too bad. A basement below an underground building, as we proved when building the gallery wing, has to be so heavily reinforced it gets very expensive.

3

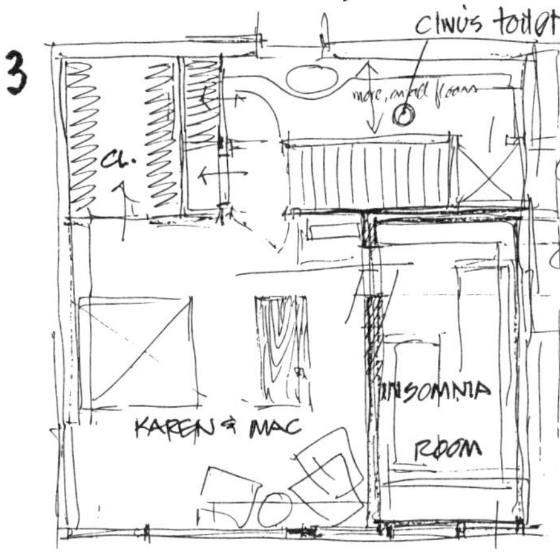

Waterproofing Underground Buildings

I knew I'd find room for it.

In my almost 20 years of actual construction experience with underground buildings I've never had (knock wood) a wall or roof leak. Skylight leaks, yes. Air leaks around projecting beams, yes, but no underground leaks.* Right from the start, I seem to have stumbled on the right combination and sequence of materials, i.e.-

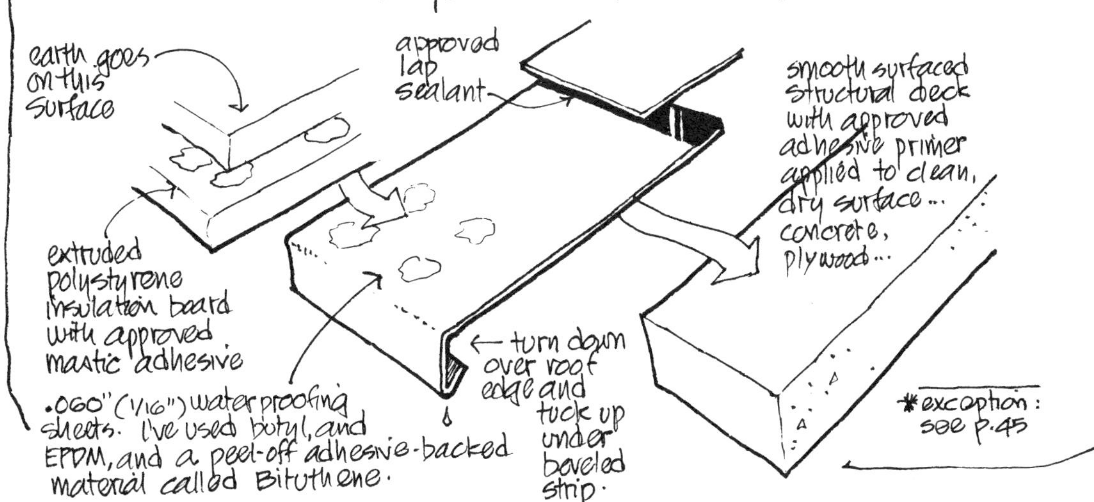

*exception: see p.45

2

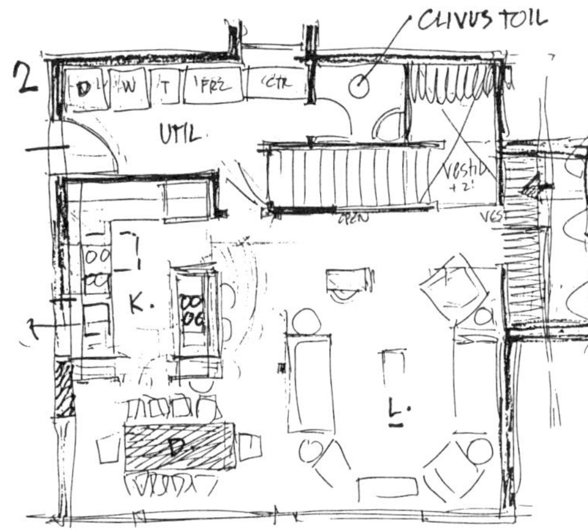

1

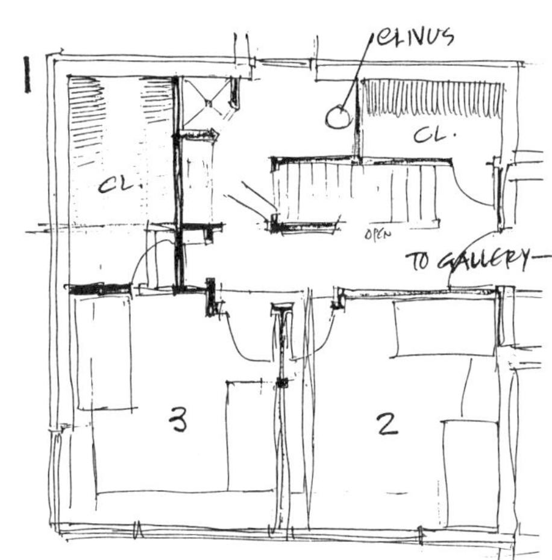

When a nearby supermarket had stood idle for several months, a local theatre group proposed that it be turned into a neighborhood playhouse. There was talk of attracting theatre crowds from as far away as Boston. 90 miles. Could be.

I couldn't resist offering this tempting earth shelter for the arriving Bostonians. A few days later, the building was rented to an exercise club, and that was that. Good exercise for me, though.

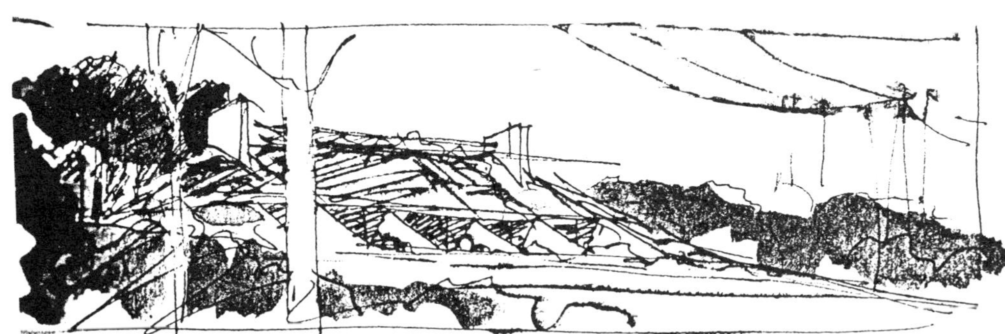

An earth house – to be built near a power line.

And then there was the developer who once let it be known that if Karen and I didn't sign a certain piece of paper he'd build a lot of low-income housing not far from us, thereby presumably lowering the value of our property.

He didn't know he was dealing with a couple of suburban liberals. We thought it sounded like a great idea rather than a threat, and I sent him this design for an earth-covered low-income housing project.

Funny. He never raised the subject again.

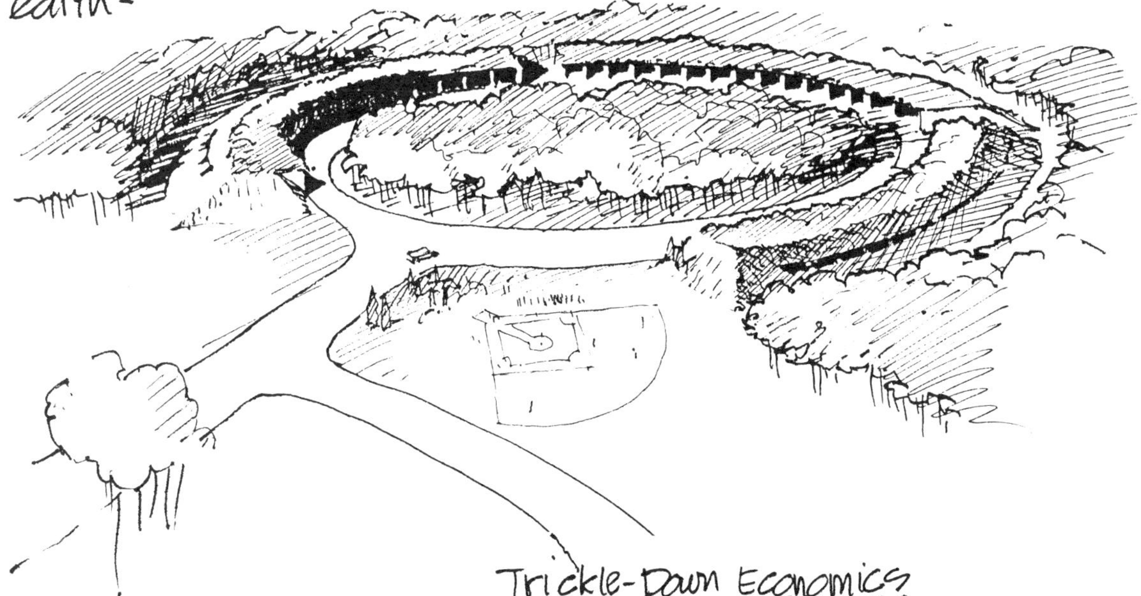

Trickle-Down Economics

It will be a long time before we see low-income housing harvesting the benefits of living in harmony with the land. My clients are usually pretty well off, financially, and my hope is that what they build will benefit less fortunate people by getting all the bugs out of the work.

Kennels are not very popular if they're built near residential areas. Most people, apparently, are like me: cranky about the sound of dogs barking in the night. Or in the daytime, for that matter.

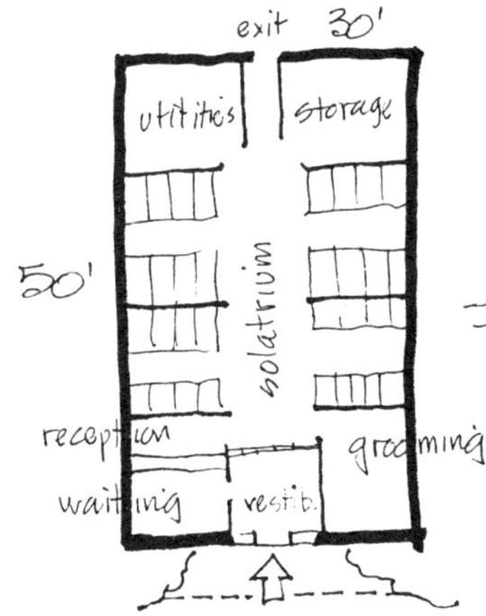

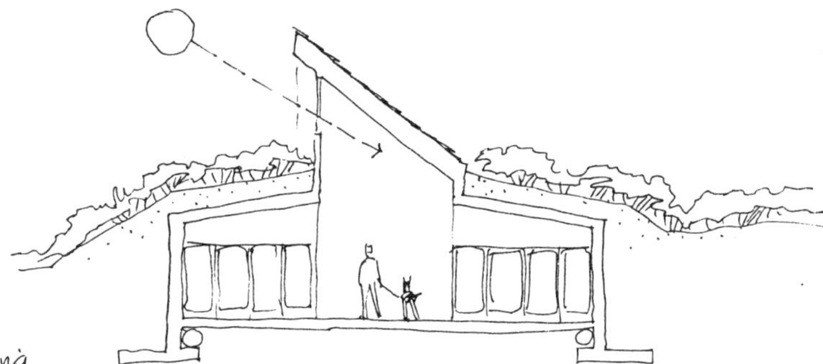

This client wanted - or, I should say, wants, for the project is still on hold - to be a good neighbor. He has picked the most remote part of his site for a kennel, and he plans not only to cover it with sound-deadening earth but also to engage an acoustic engineer to make sure that annoying sounds don't escape through vents or open windows.

I picture a great sun-filled central space and a lot of happy dogs, secure in the knowledge that man has at last come to understand that there is in all canines an atavistic memory of life in dugouts and dens.

I'm encouraged by memories of my office at Cherry Hill, New Jersey, where the noise of the nearby traffic was somehow lost in the open courtyard set 7' below grade level. It wasn't silent by any means but it was surprisingly quiet.

Not long ago I discovered why they print all those drawings of new and used cars in the back section of the newspaper, after the classifieds and the sports. The drawings are there, apparently, for architects to cut out, or to trace, for use on their building perspectives. The only trouble is the vehicles then look more convincing than the buildings.

A design for John Hait.

A mower sales & service shop in NY state.

The majority of American underground houses may very well look like this

but I've seen this one somewhere ... in Wisconsin or Minnesota

and in Kansas or Iowa this treated plywood arrangement turned up, with two trees as the only hints of the existence of rooftop earth.

For a while I toyed with the idea of having sun reflectors outside windows in the winter. But the problems...!

Better to stick with what I know will work, and the sniffer duct + underfloor tubes combination works really well on very little electric power.

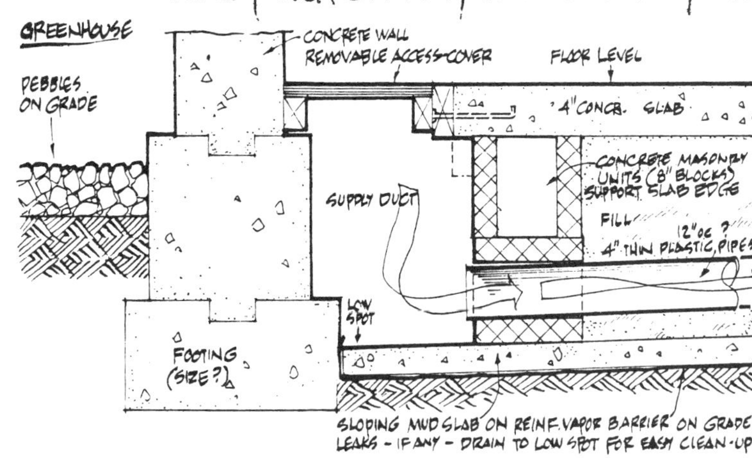

A related detail appears on p. 137.

John Hait (see p. 48) is the man who knows about using "umbrellas" of insulation in the earth to hold great amounts of warmth in the surrounding soil. Before I read his book I'd heard rumors about the practice, leading me to produce this variation on the umbrella scheme. This scheme misses the best, the warmest place, to store heat: right above the roof. But then that's where the most severe weather exposure occurs. So....?

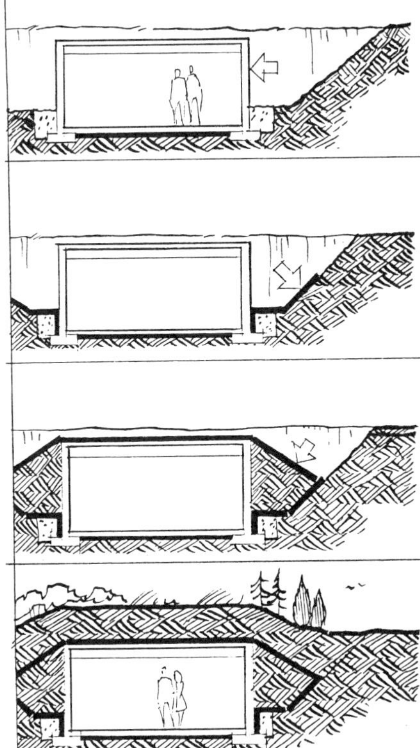

Structure is insulated on lower surface, and waterproofed (arrow), ready for backfill.

Insulation board (arrow) is applied to carefully-smoothed bank, up to halfway line.

Backfill is placed and tamped around building, insulation board (arrow) is applied to carefully-smoothed slope.

Backfill, topsoil, mulch, and landscaping are added. Earth "inside" insulation stabilizes building temperatures.

reflectors and sun control . . . .

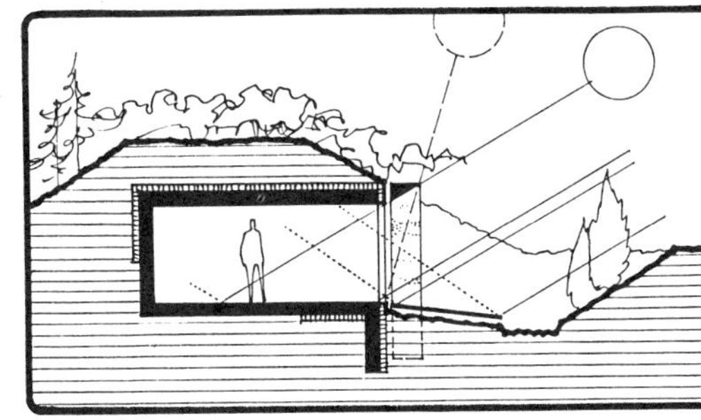

plus cross ventilation and more sunlight . . . .

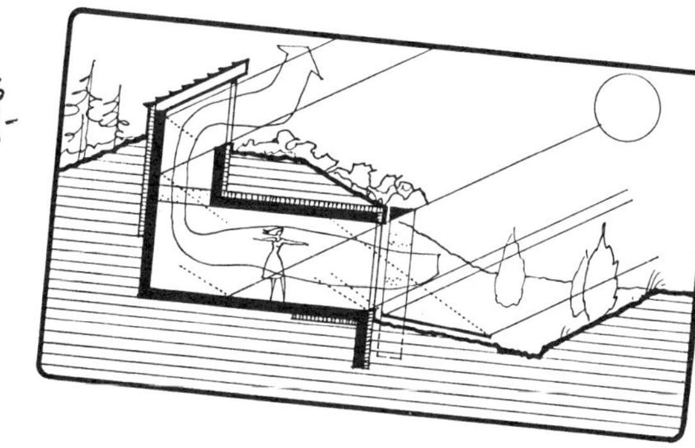

with a sniffer duct system added . . .

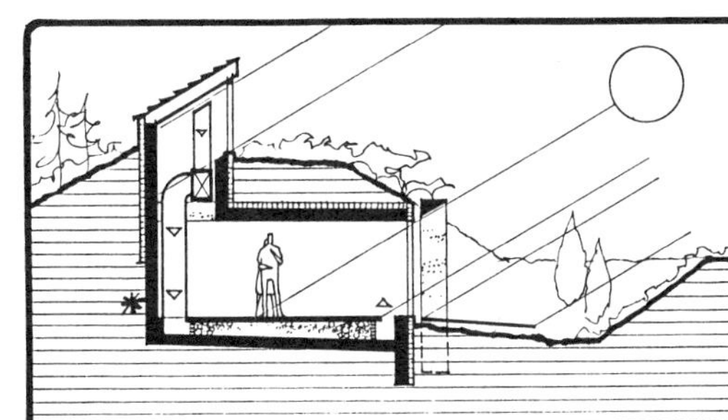

and multiplied by 200 equals Happy Hill Acres.

*see that asterisk? Think of all the heat washed away from the wall as rainwater, percolating downward, steals it.

It could be the entrance to a turnpike of the future but it was designed to be the entrance & control point of a big new golf course.

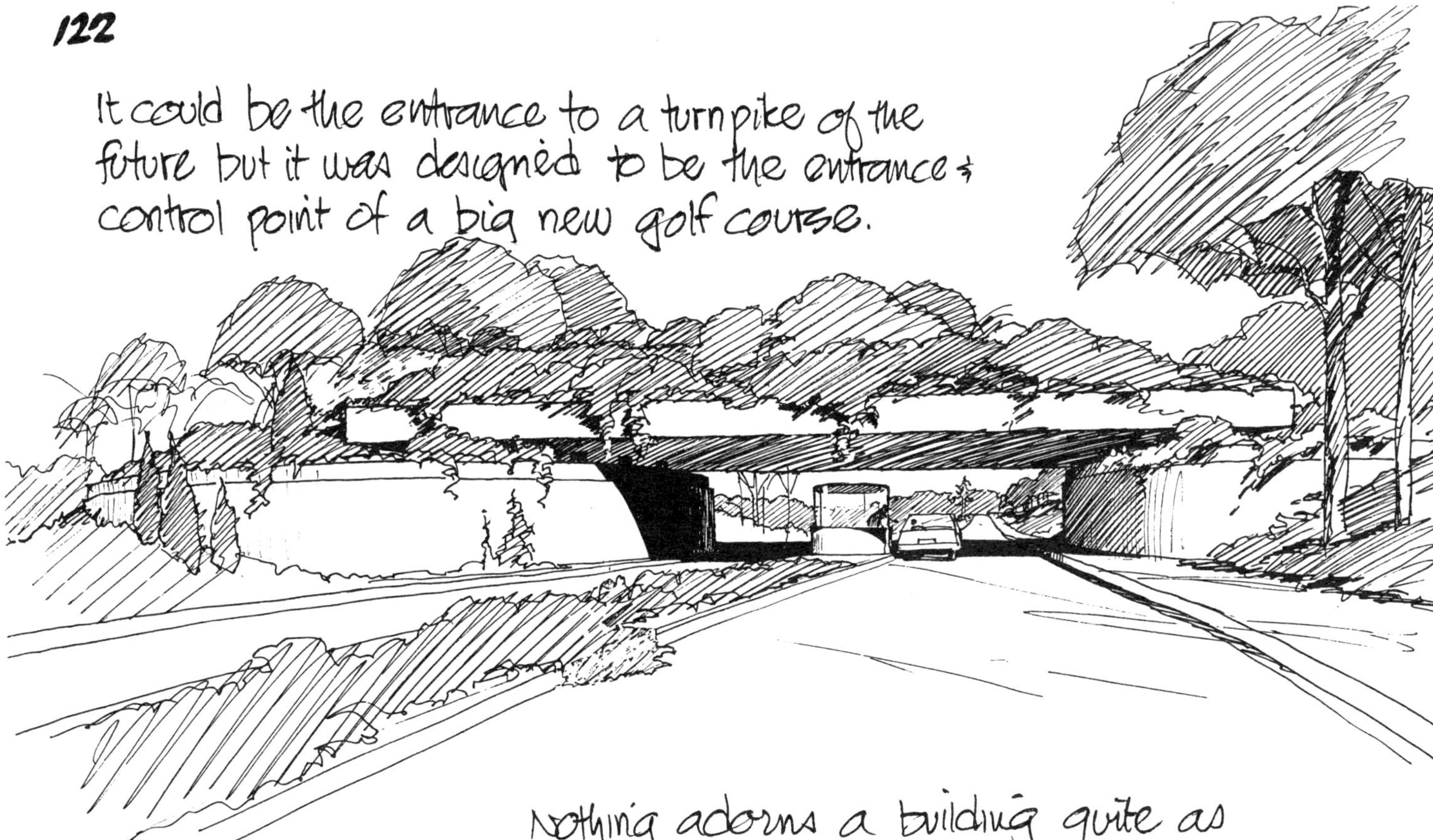

Nothing adorns a building quite as beautifully as do masses of plants. If they feel uncomfortable with the way they were arranged they will rearrange themselves so naturally that the result is always far better than any human artist could achieve.

A little farther in, along a winding drive, a straightaway appears, and there, at the side of it, is the clubhouse itself with its 3 tiers overlooking the course on the other side. I'd be so happy to see a golf club build something like this I'd be almost tempted to try playing a game just to see what it is that people like about hitting a little ball for a mile or so.

This was not a make-believe project. In 1987, I was asked by the club's trustees to submit a design for the clubhouse they hoped to build when all the paperwork had been done to permit construction of this course in southern New Jersey.

"A very interesting concept," they responded, not quite killing all my hopes.

Well, someday, somewhere, it's bound to happen.

I'm perfectly willing to outwait them, for goodness and truth are on my side. Why don't they recognize that?

This little earthen skywalk is an embellishment taken from some of my correspondence with the club trustees.

Who could fail to resist the charms of such a structure?

Let me just whip out a little structural sketch. Waterproofing sheets can be cut and folded to cover almost any combination of flat surfaces, especially when complications like pipe penetrations are not involved.

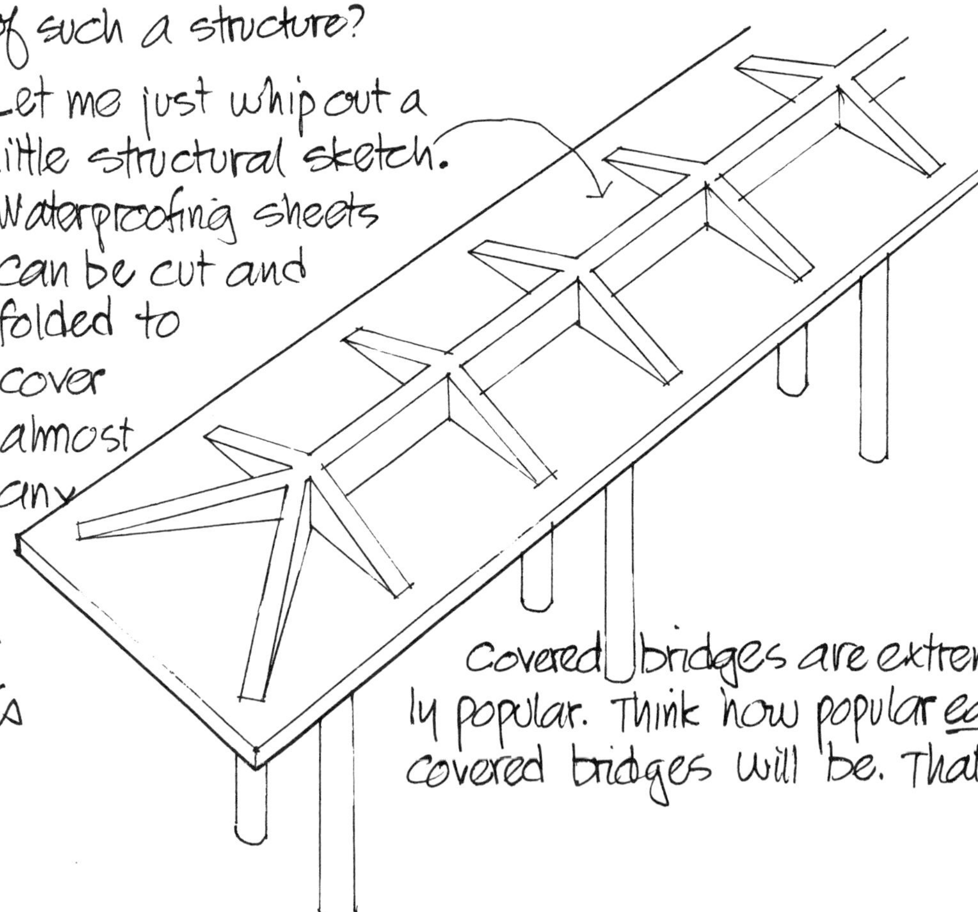

Covered bridges are extremely popular. Think how popular <u>earth</u> covered bridges will be. That's why, when I was asked to design this footbridge into a new tri-level office building, I never thought twice about the earth-covered part. It's the only way to go.

(Still "on hold", this one, also in New Jersey.)

By the year 2050, there will be hardly anyone still alive who remembers when there was even a dispute over whether living plants should be made the central design feature of architecture. For both practical and impractical reasons, it's simply the best way to treat outdoor surfaces on this particular planet.

PLANTED ROOFTOPS (BEYOND) TEND TO DISAPPEAR WHEN VIEWED FROM PRIVATE PATIO-BALCONY

China! 1985.
An earth-covered, solar, organic health center in China!

I got so excited when I heard about the project I could think of little else for weeks. The cost of the books alone was more than I'd paid for books in years. China was my life, and I slowly came to recognize the names of places that, a few weeks earlier, had been utterly unknown to me. I was going to China! Money would be forthcoming. It was all set.

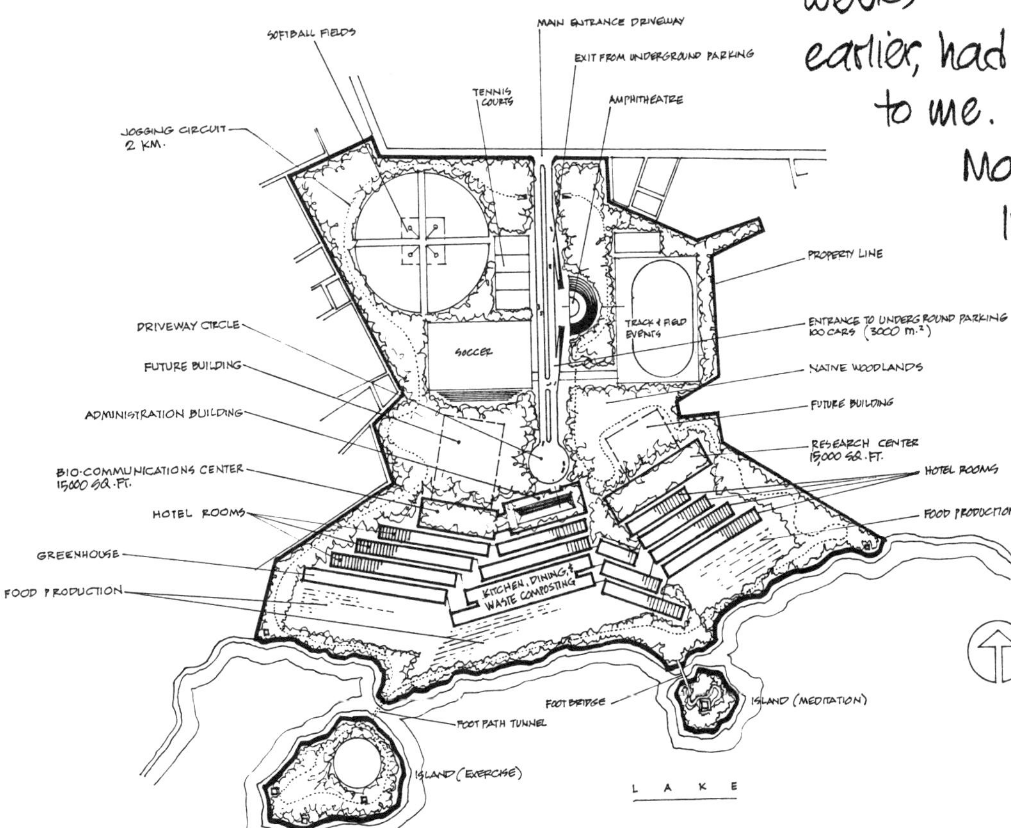

I worked 12, 14 hours a day on the designs, making sure the rooftops really would disappear as, in the caption above, I had said they would.

But then the client disappeared before the rooftops ever got their chance to do the same.

It's not the unpaid bills that hurt so much. And I'm not at all sure, now, that a suburban U.S. architect should have been involved there. I just don't understand how that slippery guy could have dropped the project without saying a word.

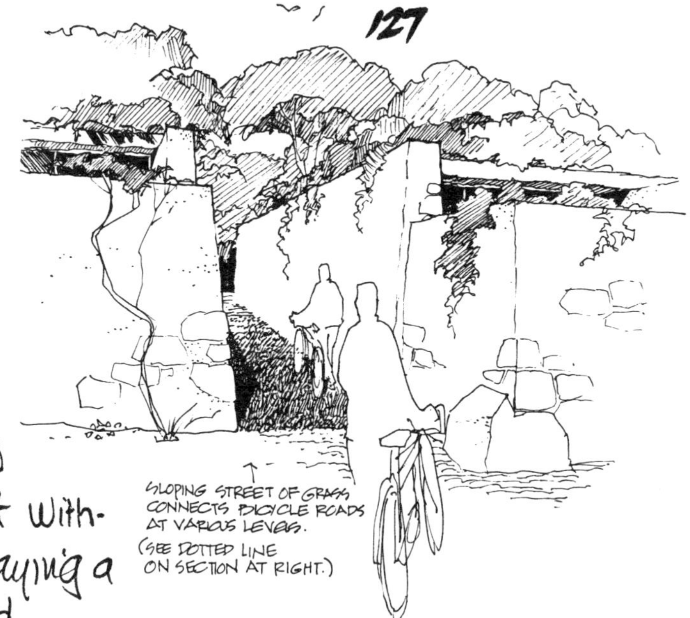

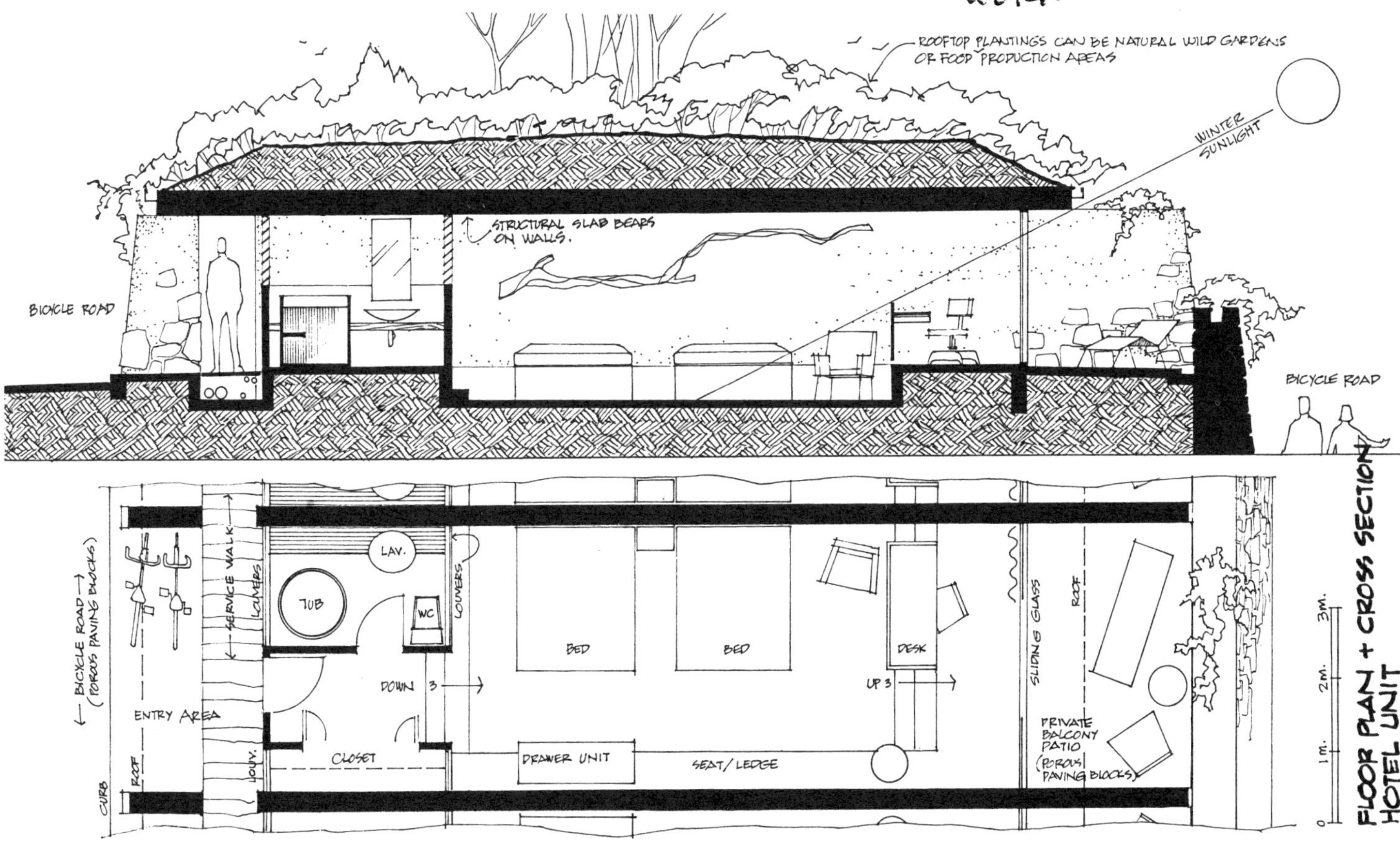

Hmm... haven't we seen this hotel room design somewhere before?

page 41, perhaps?

Northwest of Philadelphia, there are places where the last glacier ran out of steam and left all its debris for someone else to pick up. 10,000 years (or so) later, some clients asked me if it was possible to build a house on a hillside covered with car-sized, and even garage-sized rocks. I wasn't sure till I saw them. What I found was a forest of big old trees growing among, and occasionally on or through, the boulders.

"Stilts," I said, and "earth cover. Passive solar, too."

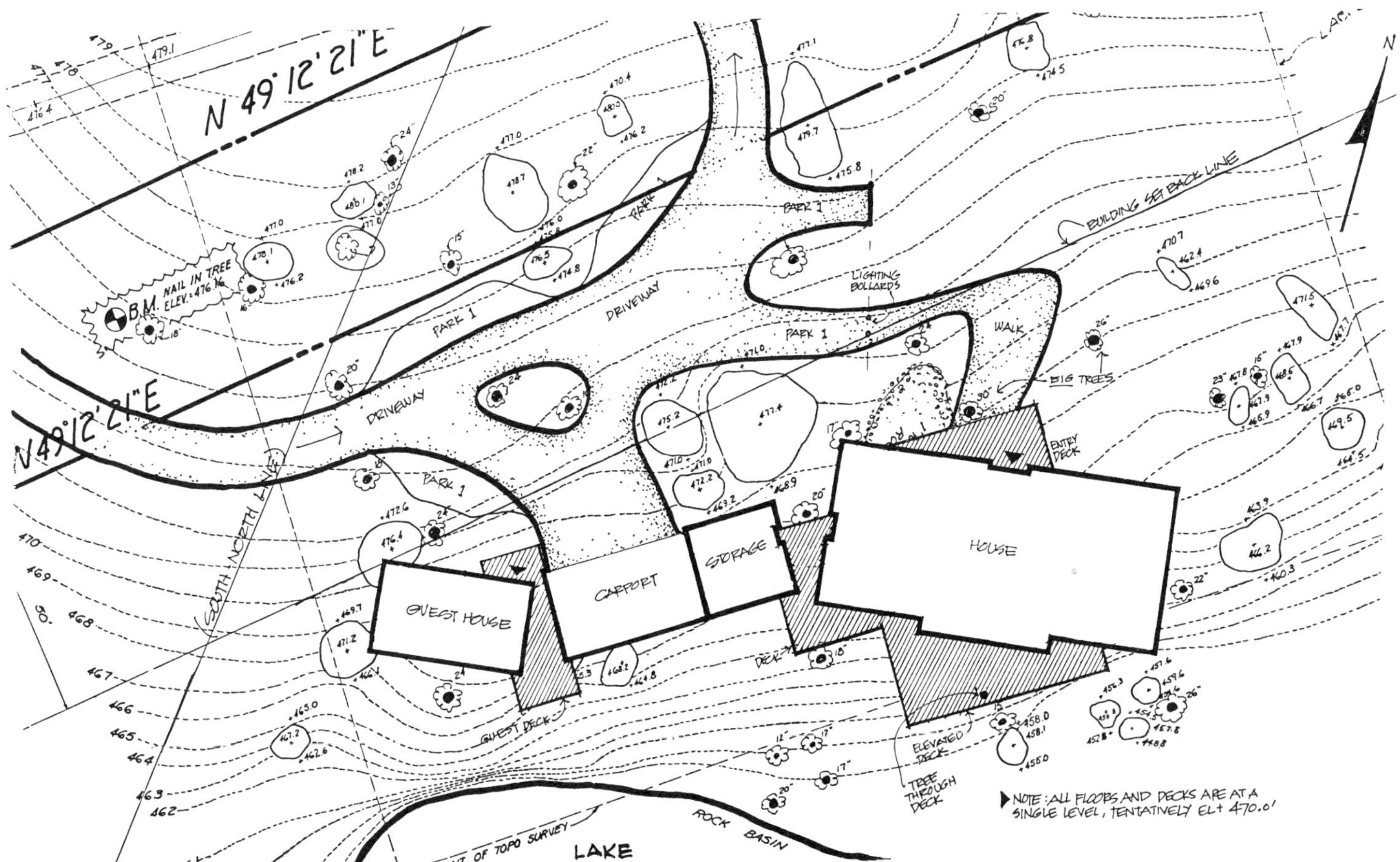

It seemed obvious from the surveyor's information that the house could go in only one place, especially if it was to have a good view of the small lake on the southern edge of the site.

An under-ground house, up in the air, on the rocks, so to speak? It seemed appropriate enough ... until the bids came in. Getting the earth up there by crane on that wooded, bouldered site — and then supporting it — strained the budget just a little too much. The owners were as reluctant as I but the earth had to go, so they built the house with an almost-flat tar-and-gravel roof.

On these pages you'll see the earth covered version for the last time.

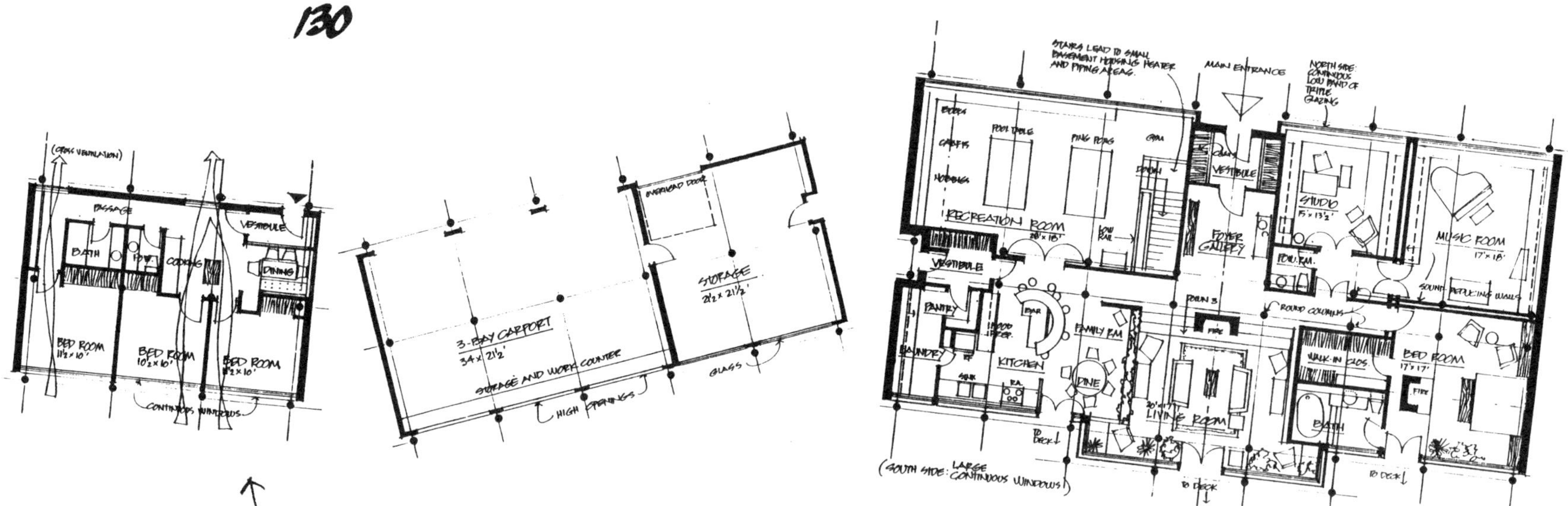

(The lakeside deck linking the structures isn't shown.)

Yes, as usual, and perhaps rightly so, the budget prevailed. The architect selected to do the construction drawings and manage the project managed to save even more money. By putting the guest quarters on top of the carport, and simplifying the plan, he got the figures to come out right so that construction could begin.

I should probably be publishing his version of the design but this is my book, and the subject is underground architecture. The question, in that regard, is: was this a justifiable use of earth cover? Even if the budget had been unlimited, would the greening of these few thousand square feet of roof surface have been defensible in any world-environment sense? I'm too biased to say. The answer has to involve the energy "costs" of placing and carrying all that weight vs. the environmental damage caused by an impervious, lifeless rooftop over the lifetime of the building.

I want to hear what the answer is after, say, 50 years have passed. Our environmental perspective should be in much sharper focus by then.

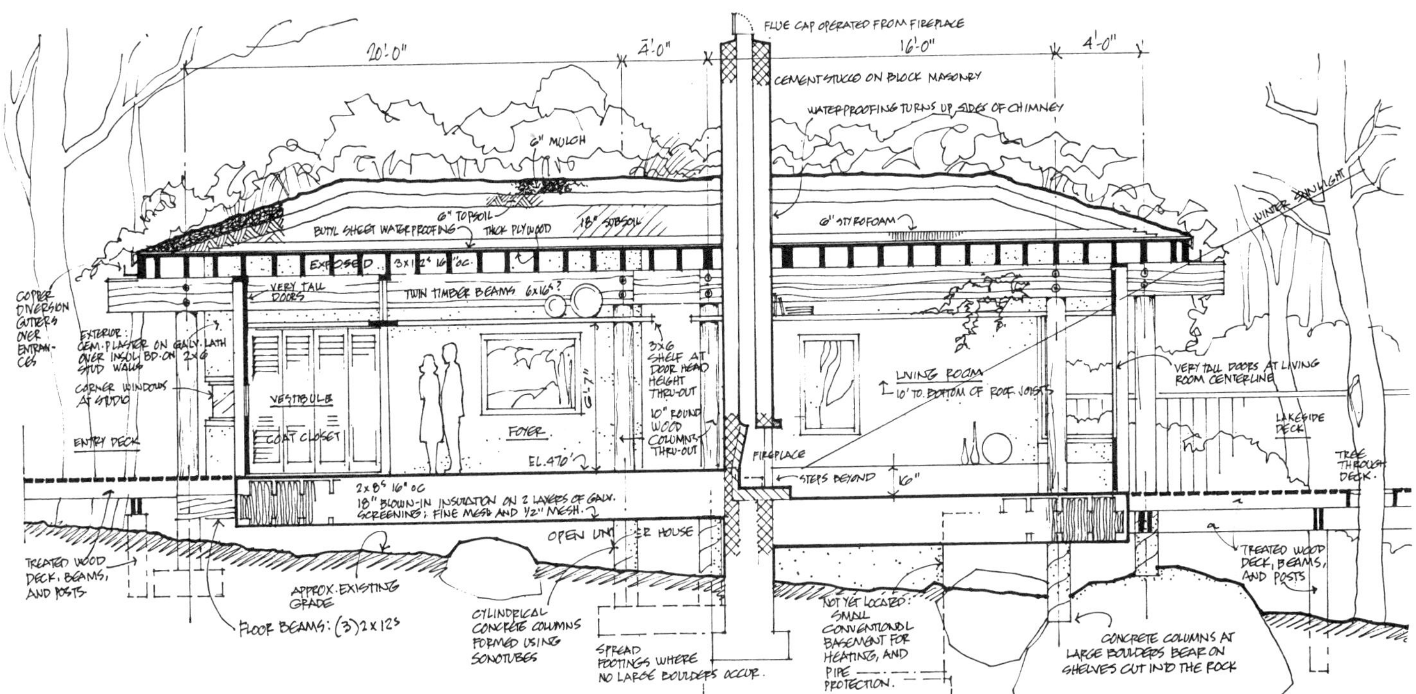

CROSS SECTION THROUGH FOYER AND LIVING ROOM

An airlock entrance vestibule on this north side, along with small, sit-down-height windows, helps resist the tendency of hard-won solar warmth to escape through to cold side of the house. Cement plaster over heavily-insulated walls completes the weather armor.

N.E. Corner

The note at the bottom says it all about this interior view, looking toward the southeast.

When wood columns are involved, I'm getting to like round ones more and more, especially when they have the natural roundnesses of tree trunks.

All those tons of earth weren't to be carried by the bolts alone. The columns were to be notched enough to offer at least partial support directly on their shoulders.

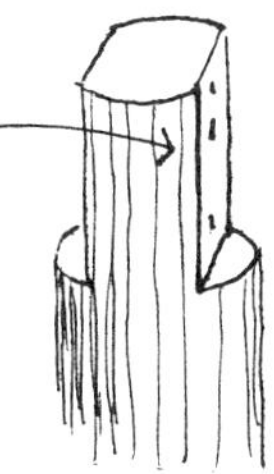

If there's confusion as to which are columns and which are trees, well, that's the way it's supposed be.

Oh, my: there are those 1950s chairs again! And here I am, still not seeing what's wrong with them.

LIVING ROOM — LOOKING WEST THROUGH THE GREENHOUSE, WHICH IS BEYOND THE GLASS PARTITION. DESPITE ABUNDANCE OF CROSS VENTILATION SUMMER SHADES WOULD BE NEEDED.

The wood truss house, from the book* my son, Sam, and I did in 1980, is one of my favorites but in 10 years no one has even mentioned it, let alone built it. I think the experience of walking among the great floor-to-ceiling structures would be a continuing delight.

---

* Underground Plans Book 1. 1980 $15 postpaid, right here.

Truss house, exterior view.
Rainwater, from the roof-edge gutter, will run down the hanging chain and step off onto the ground without a splash. It's an old, old - colonial or earlier - device I borrowed for your pleasure.

Another house from the same father-and-son book, this one aroused just about as much interest as the truss house.

More details from *Underground Plans Book 1*, the book I saw, in the heat of America's energy-crisis infatuation with solar/earth shelters, as but the first of a great series. Sales were brisk for a while, and orders still come along at the rate of 1 or 2 a month. Lately, there's even been a slight and consistent upturn in sales, and it was on the basis of that upturn that I gathered my courage to publish this book. With oil prices still relatively low, I can attribute the recent upturn to only one thing: an emerging environmental concern. And that's the best news of the decade.

IF EARTH AT ROOF EDGES IS RAKED BACK TO ITS NATURAL ANGLE OF REPOSE AND THEN COVERED WITH DEEP MULCH THE BANKS SHOULD NOT ERODE. IN DRYER PARTS OF THE COUNTRY (NOT DESERT AREAS; WE DON'T KNOW WHAT'S BEST THERE) BALED HAY "WALLS" ALONG ROOF EDGES, WEIGHTED WITH ROCKS, AND SURROUNDING LEVEL SOIL BEDS, SHOULD GIVE NATIVE GRASSES AND WILDFLOWERS A CHANCE TO TAKE OVER AND HOLD THE SOIL.

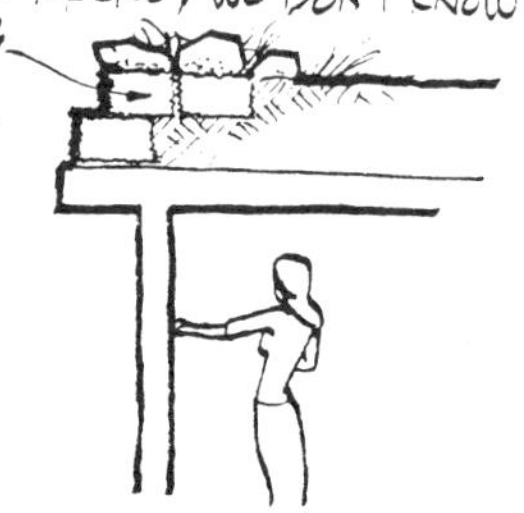

REMEMBER THAT UNWARY PERSONS COULD FALL FROM THE EDGES OF EARTH-COVERED ROOFS. YOU'VE GOT TO FENCE OR IN SOME OTHER WAY DISCOURAGE ROOFTOP TRAFFIC. LOW-COST WIRE FENCING, PAINTED A BROWNISH GREEN (OR WHATEVER THE AVERAGE YEAR-ROUND COLOR OF THE LANDSCAPE IS IN YOUR AREA) AND ZIG-ZAGGED AMONG THE ROOFTOP PLANTS, CAN BE BOTH EFFECTIVE AND INCONSPICUOUS.

OPEN SHEDS, FACING SOUTH, FOR THE STORAGE OF LUMBER, FIREWOOD, CEMENT, SAND, FERTILIZERS, MAINTENANCE EQUIPMENT, BOAT, CAR, MULCH... AND PERHAPS FOR AN OPEN-AIR WOODSTOVE OR FIREPLACE – IF THE "SHED" IS NEAR THE GARDEN – FOR CANNING AND SEASONAL COOKING. A WARM SUNNY SHELTER LIKE THIS, EVEN ON A BITTER COLD DAY, CAN BE A GREAT PLACE TO COOK AND SERVE A MEAL – IF YOU'RE DRESSED PROPERLY.

HOW ABOUT A SCREENED PORCH, PERHAPS A DETACHED ONE... ON THE SAME LEVEL, FOR EASY ROLLING OF SERVING CARTS? WHATEVER YOU DO, BE SURE TO LET THE STRUCTURE ECHO THE DESIGN OF YOUR HOUSE. IF THERE ARE NO BUGS WHERE YOU LIVE YOU CAN CALL IT A SUNSHELTER OR A RAINHOUSE.

There's very little, if anything, to add to what these details have to say so I'm going to go have a cup of coffee while you study this page.

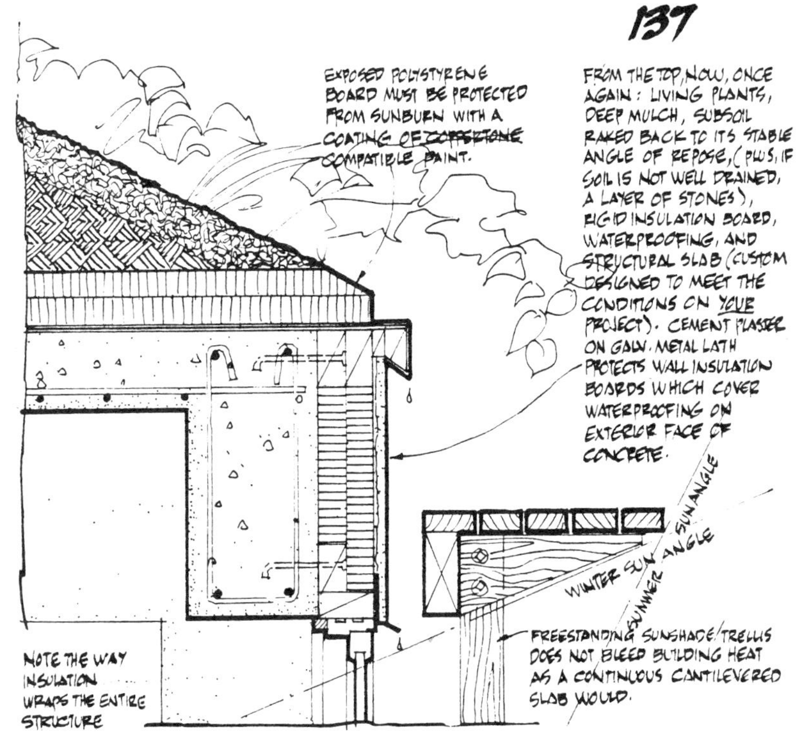

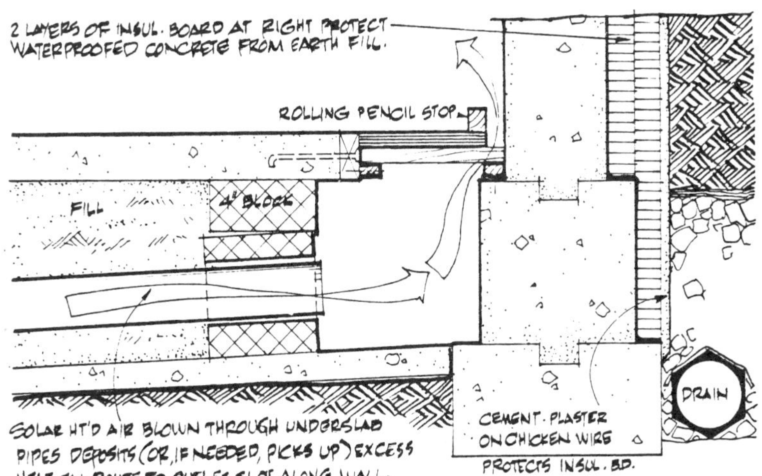

← (The companion to this detail can be found on page 120.)

In his infinite wisdom, the editor of Popular Science decided to publish this house, called "Low and Bermed," along with another of my design (a house on stilts for construction in low areas), in the June, 1984, issue. Readers were told that if they sent me $63 they would be sent construction plans, details, and specifications for the house they preferred.

If I had plotted the sales, they would have fallen, quite predictabley, on a graph something like this: 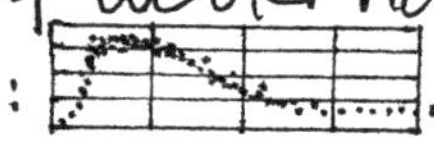. What I didn't predict was the length of the tail on that graph. Here I am, six years later, still getting occasional orders for the houses. The latest was an order for the stilt house drawings. It came from Pago-pago!

←This house, however, is the sales leader.

The many uses to which the smaller of the two buildings can be put appears to be responsible for a good deal of the enthusiasm for this design. Storage room, workshop, hideaway, playroom, the 14'x12' structure can be modified quite easily for use as a guest house or even a garage.

Steep berms, echoing the slopes of the roofs, offer weather buffering, a landscape theme, and a unifying touch.

Can you identify all the rooms?

The house is 28'x36'.

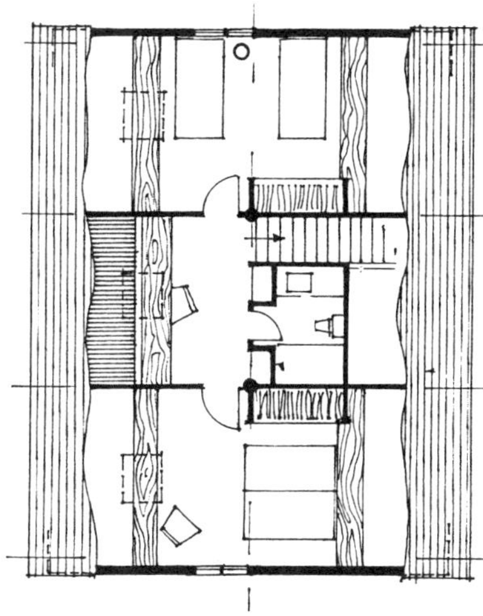

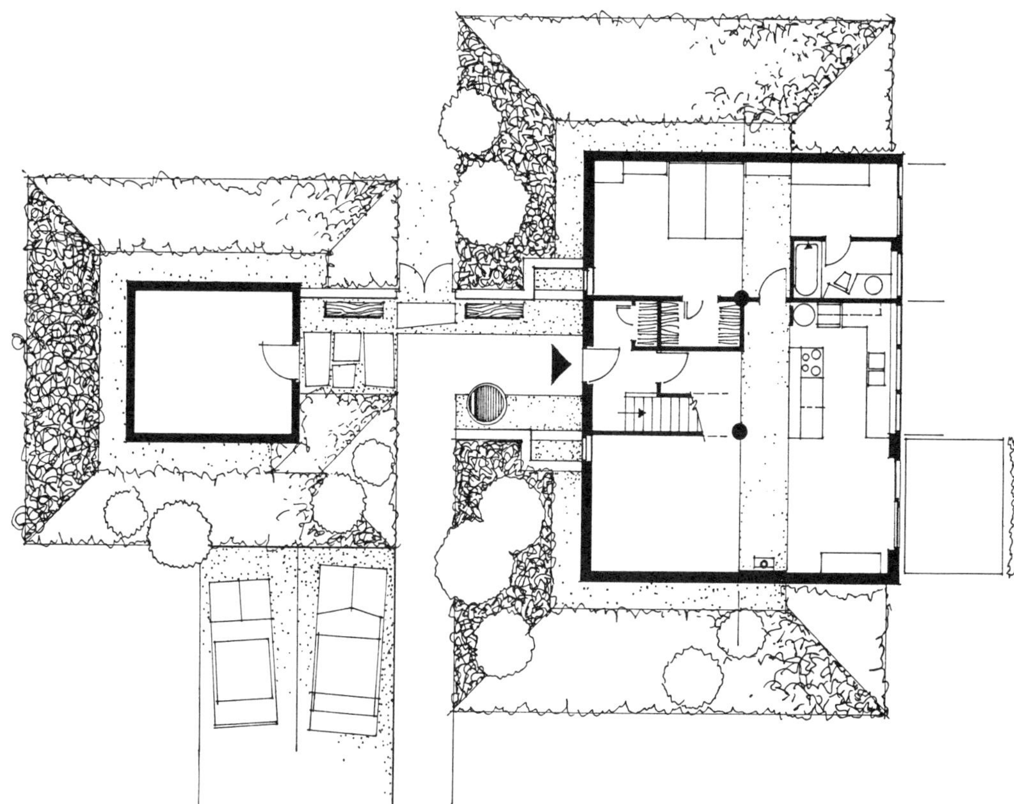

Dogged persistence, isn't that the key? How many times have you seen this pond-at-the-window-sill design so far? Two? Three? I just keep pushing for it, knowing that sooner or later it will get built.

The key to its success, of course, is in having an absolutely fail-safe drain, by gravity, from the "safety overflow and maintenance walk" between the water and the building.

First-time visitors will no doubt be just a little uneasy in a room so obviously below water level but, after a time, the nearness of ducks and boats, of ice or ripples, and of magnificent reflections should be downright addictive.

By holding back the supporting columns, the wall of glass is left nearly unbroken, offering sweeping views of the lake.

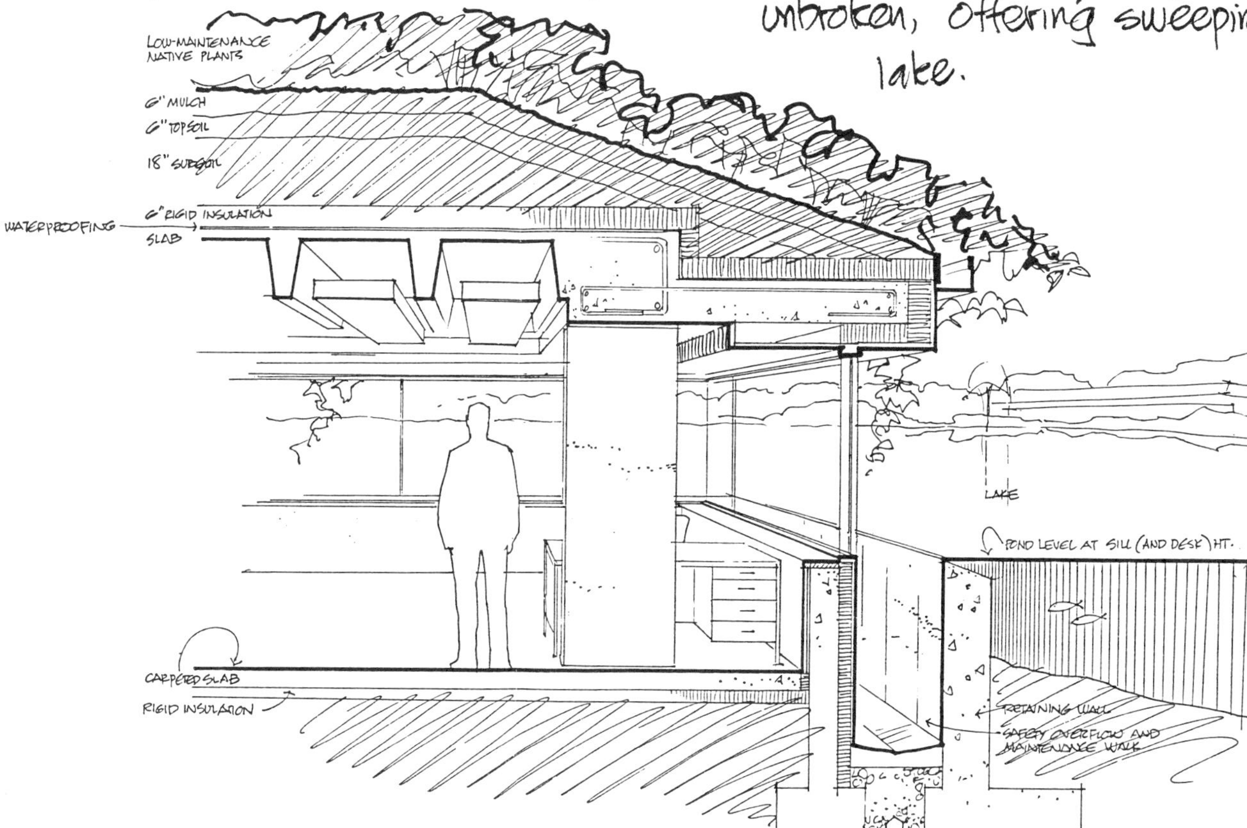

Every time I refine this idea I wonder about the surprises that lie locked up in it somewhere, waiting to cause trouble. A tidal wave? Not on a little lake. An earthquake? What damage would it do? A 500-foot oil tanker gone far astray in the fog? The Loch Ness monster? In my experience so far, the few surprises – negative ones – related to earth cover have been far out-weighed by the positive ones. (May it always be so, he wishes with fervor.)

"Send us a proposal," said the fire station's building committee. I did, and illustrated it with my recommendation. They were never heard from again, either. No doubt they'll build another Cape Cod cottage with garage doors across its front. Ah, but what a joy it was to dream this dream of an underground fire station!

Right at the edge of a cranberry bog, the site is certainly one of the most idyllic of American firehouse locations. Seven years have passed so far, with no sign of activity, not even a sign of a construction sign on the site. I go by about once a month on other business, never failing to check for progress there.

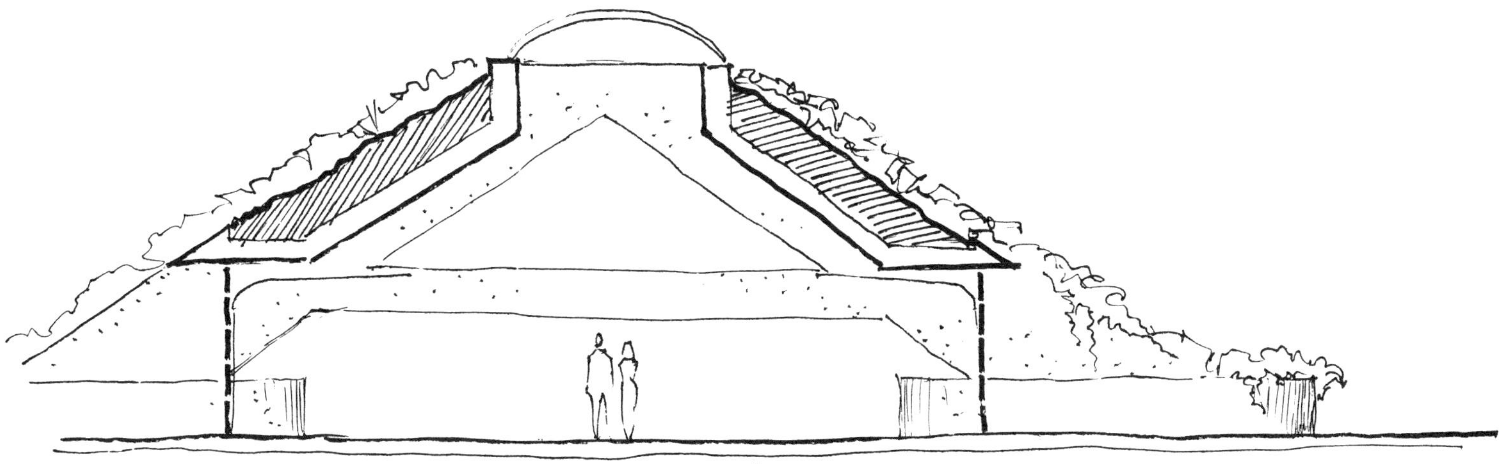

The great skylighted space, the drive-through ease of moving the great engines, the flowering hillsides, and the fireproof construction ... even the potential for having offices, storage, and bunk rooms up on a mezzanine level .... how could they resist such an offering?

Only with great courage, I'd say.

If you'll turn the page you'll see even more of the delights so manfully resisted by these Cape Codders.

HARWICH
69
Malcolm Wells

Earlier, and for another fire house project, I'd done this smaller, double-doored design. Between the doors ran the east-west corridor, fluctuating in temperature because of its single-door protection, but inside the bays a stable and comfortable environment would have been easy to maintain.

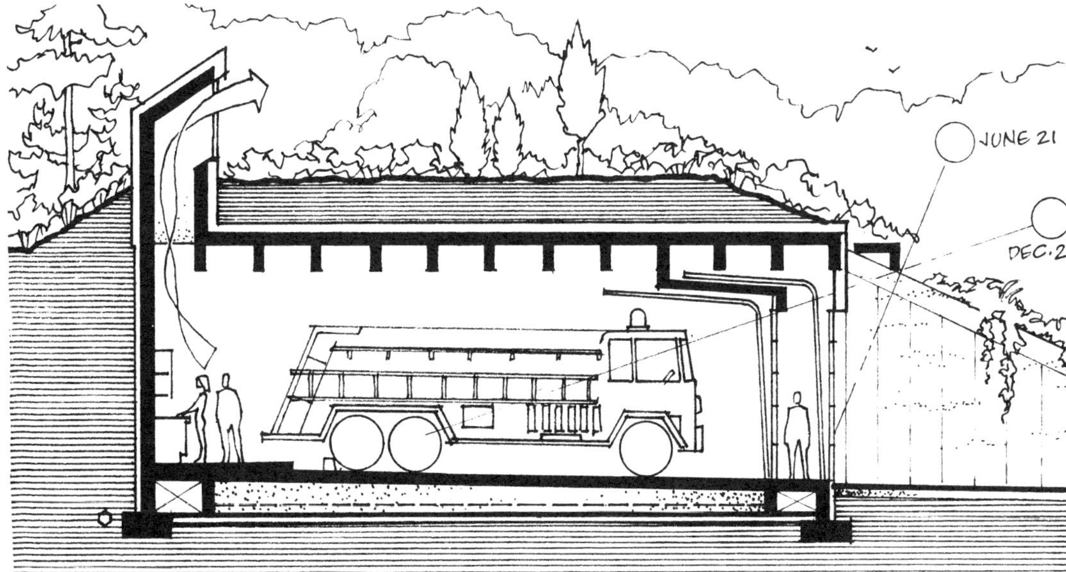

SCHEMATIC CROSS SECTION @ 1/10" SCALE

NOTE UNDERFLOOR THERMAL MASS, SOLAR HEATING, CORRIDOR THROUGH AIRLOCK SPACE BETWEEN GLASS OVERHEAD DOORS, NATURAL THROUGH-VENTILATION, AND NATIVE PLANTS. NATURAL LIGHT. EXTERIOR INSULATION LETS ENTIRE BUILDING STORE HEAT. COOL IN SUMMER WITHOUT CONDENSATION ON WALLS AND FLOOR.

So, which would you rather have, an earth-covered firehouse or a window sill precisely at pond level? It's a tough choice, isn't it?

"First Light", it was to have been called ... a giant theatrical complex in the forest by the sea, here on Cape Cod ... a place where the first encounter between Europeans and Native Americans could be presented, with dignity, drama, and authenticity in an outdoor setting.

The true story was to be told from the Indians' point of view. Historical consultants and some of the actors were to be Cape Cod's Wampanoag Indians. A reproduction Mayflower would sail across the outdoor "stage". The woods themselves would be the backdrop.

Porous paving and an earth-covered roof were to minimize the land damage. Environmental rectitude was to be the design theme. Public interest was aroused, the news media were helpful, and hopes were high. But as site selection difficulties mounted and time began to drag, interest began to drag as well, and the entire project was put on what may be a very permanent kind of hold.

←You can probably tell that I winged that perspective, freehand, without laying out vanishing points. It's probably a little distorted as a result but I think it captures the spirit of the entire design. R.I.P.

Remember On The Level (p. 52), the book that almost made it to the printer's? This cartoon was to have been a part of it. Shaking the idea that we must impress our neighbors with a facade that has meaning to them is by no means as easy as it might seem. We – and they – rate conformity right up there with purchase price when it comes to the criteria for selecting a house. We don't admit it, of course, but we all want approval and acceptance, if not outright admiration, from our neighbors.

I couldn't resist putting the Sidwells' name on this drawing because they, of all people, are probably the least concerned about curbside appeal of any couple I know. You must remember their house, back on page 88.

The all-on-one-level book idea came about as a result of requests from all over the country for some designs that would be appropriate in flat land.

One such request came from Richmond. It was a challenge to design a solar/underground house on a vacant lot, 185' x 31', between existing brick houses. The proposed house was to have an insurance office in the basement.

Here's the result.

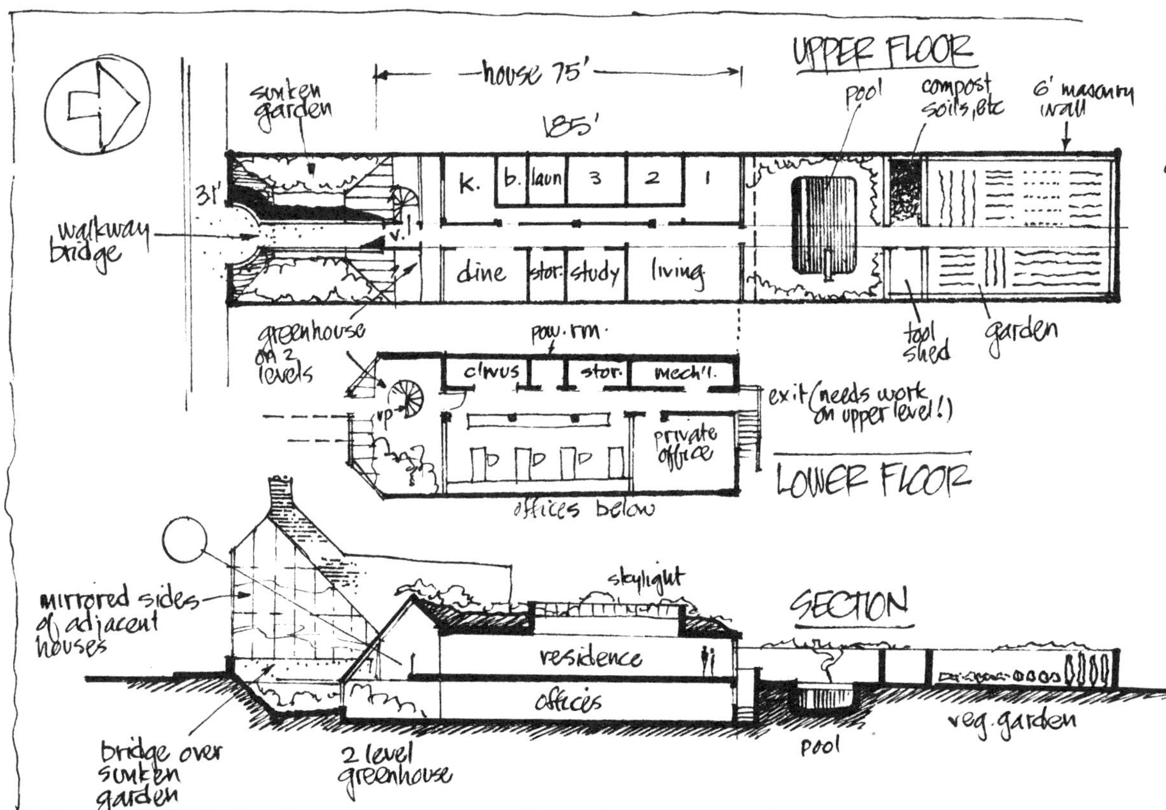

The only way I could see to get adequate sunlight into that slot between the buildings was to mirror the adjoining party walls. Would that have been allowed?

A covered (earth-covered, that is) foot-bridge from the roadside parking area to the house struck me as the key to this design. The road was on one hilltop, the house on another.

A later topographic survey, however, killed the footbridge idea; the actual grades didn't favor it at all, and back to the drawing board I went.

I had to go back a second time when estimates for the construction of the house began to reach well beyond the owners' budget. The original house (plan, next page) had grown much larger, and the basement workshop (perspective, next page) had been moved into a separate, and not inexpensive, building.

If there are two people on Cape Cod who can match the tolerance and good humor of my Raven Rocks friends in Ohio, it is this couple, Gail Turner and Peter Spier, who, unable to build their expensive dream house, asked me

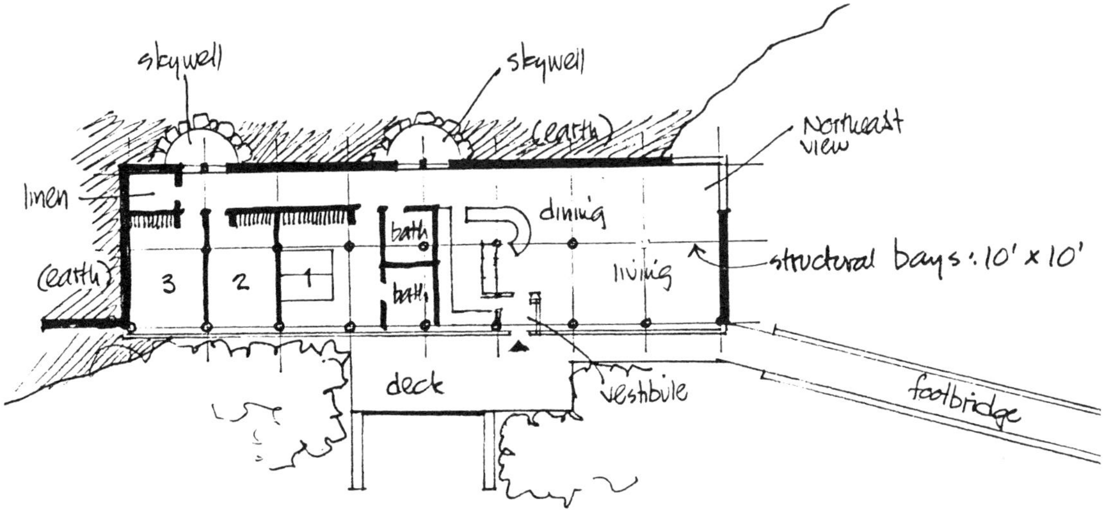

to take a third cut at the problem. This time, it managed to fit the budget, and an efficient, two-story house is now nearing completion on the site. Maybe you'll see it in the next book, but the subject here, you'll recall, is earth shelter.

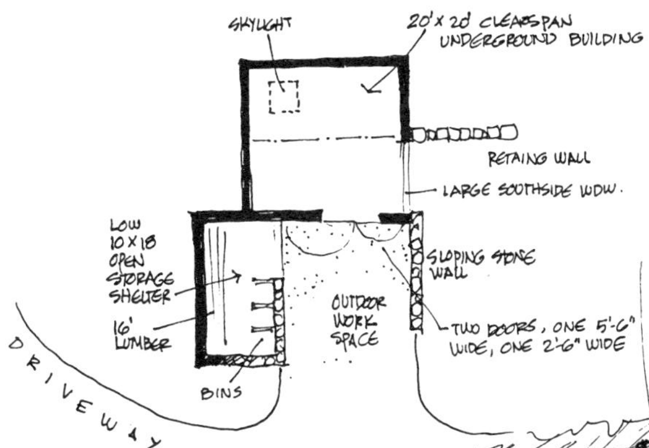

TO BRING DOWN THE SCALE OF THIS SIDE WING, I THOUGHT (1) IT COULD HAVE A FLAT SLAB ROOF JUST 7 FT. ABOVE ITS FLOOR, AND (2) THE FLOOR COULD BE SUNKEN 12", DOWN A RAMP, MAKING THE ENTRANCE OPENING 6' HIGH.

DRAWINGS NOT TO SCALE
WORKSHOP

I like to walk an owner through the design process. This stroll was taken with a good-humored school-master who had a dream of getting out of town into the lovely Indiana countryside.

1) The building(s) should be built into the south face (shaded) of the "natural ridge" part of the property.

> You mean to say we hired a consultant to tell us *that*?

2) The building(s) should be more or less linear, exposing its long face to the south.

3) A straight rectangular box would be appalling. 

4) A free-form shape would be too expensive. 

5) A rectangular (90°) plan theme would be least expensive among the irregular-plan choices but it would probably lose too much winter sunlight by casting shadows on itself much of the day.

6) The collision of two rectangular geometries is probably the next best bet in terms of economics.

(Maybe even flatter than that; like this: )

7) Most classrooms should be on the south side. This means single-loaded corridors.

8) It also means mostly earth-covered corridors, but

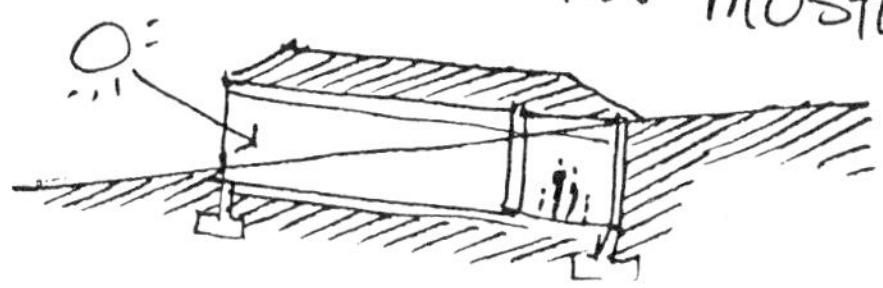

corridors must not be dark, airless, electrically-lighted tunnels, so...

9) corridors must do two things:

a) Have high windows, where possible, and
b) Poke out into daylighted sunken gardens at the ends of long runs.

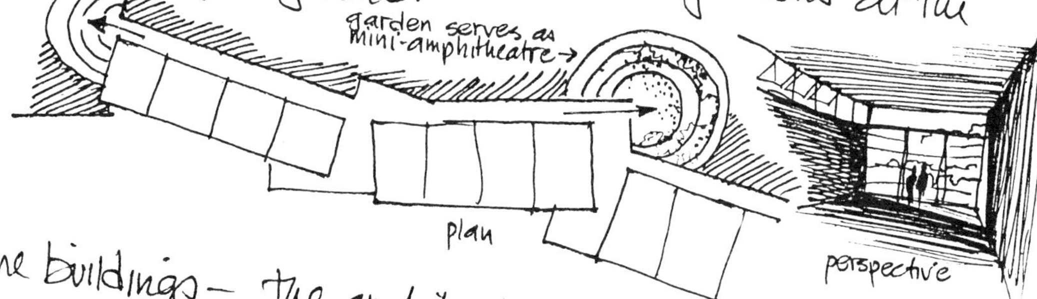

10) The buildings — the architecture — must at a glance express the idea of land husbandry, not land domination. Just as contour plowing expresses this in agriculture so does contour architecture in construction. A jumbled series of earth shelters would broadcast confused

messages. But you want varied floor levels and different ceiling heights. So do I. Such considerations, to me, lead to one conclusion: a powerful

He liked what he saw of these designs but a misunderstanding as to who was to pay for them got in the way and upset that delicate balance of trust so important to the launching of projects.

(more)→

11) you can see that I'm also suggesting the other thing that makes for a strong, simple, understandable, and economical architecture: a repetitive structural module, or skeleton, on which to hang it all.

12) I see the basic cross section as a simple one, too... Floor heights varied, roof level fixed.

13) The structure? Heavy timber? Laminated beams carrying rough-sawn joists & plank? Round natural local hardwood tree trunk columns?

are you reading all this?

14) And what about having some great, roughly rectangular slabs of your limestone standing inside some of the south-facing glass, to absorb solar energy and to reradiate it, stabilizing interior temperatures?

15) The best I've been able to do, with regard to an actual floor plan, so far, is this:

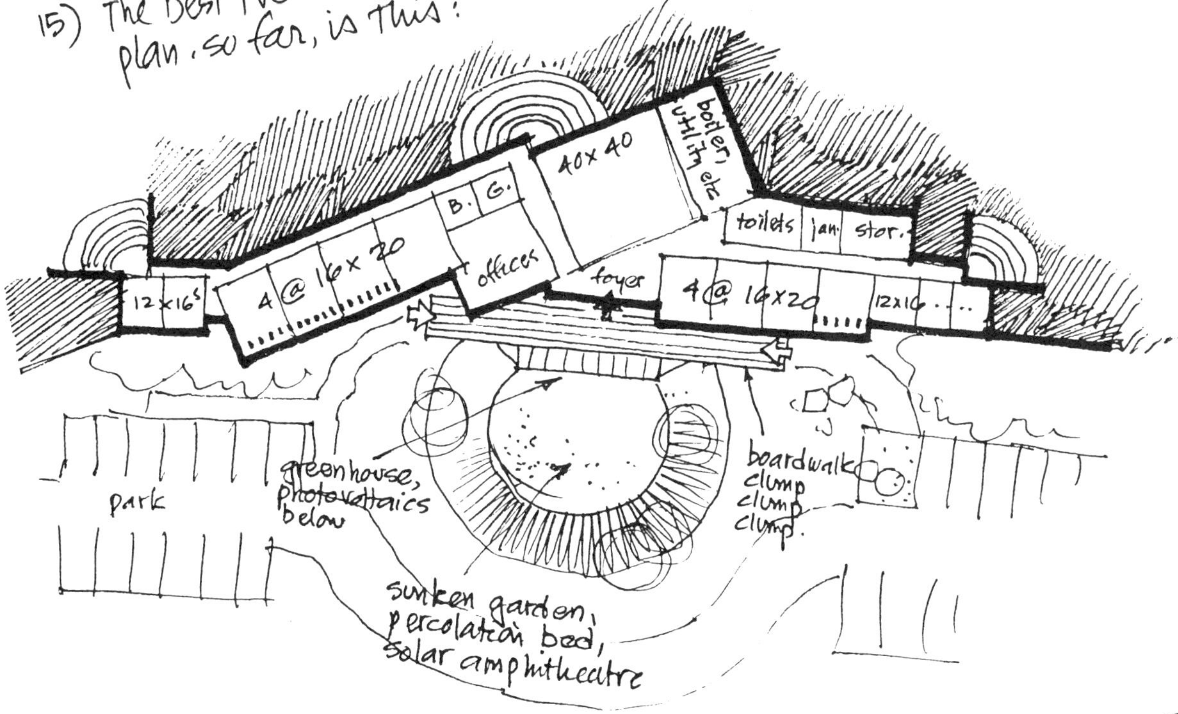

. . . and once again off into the sunset we sail.

Bye bye.

Another day, another school, this one World College West, the still-struggling little school in Marin County, north of San Francisco. Stressing environmental studies and international affairs, it seems to have just the right recipe for our times.

Founded by my longtime friend, Dick Gray, the college has managed to get itself accredited, to outgrow its original rented army barracks, to acquire a big hilltop campus, and to get some new buildings built. This, I'm afraid, is not one of them.

I was lucky enough to have been involved in the design discussions but my suggestions, as by now you can imagine without being told, were too expensive for the fledgling institution.

Above, a dormitory and amphitheatre. At left, an administration building.

You see no cars in the drawings because the parking lot is down at the foot of the great sprawling hill, near the highway. Only service vehicles and a student shuttle bus make the climb to the campus. The first time I made the climb I was so winded I thought I was a goner. After my heart stopped thudding I realized I'd better get more exercise, and I've been a runner* ever since. So you can see I got an education at the college.

* a mile or so, at a 9-minute rate, every other day, if you can call that running.

Before my first visit to Cape Cod I thought it was a sand spit so low it was probably washed over by the sea during spring tides. The forest was my first surprise, the hills were the next. None are higher than about 100 feet but they can be quite steep, as the contours at right indicate, in spite of the mostly sandy soil.

Only with the permission of the local Conservation Commission can you disturb anything within 100 feet (dotted line) of ponds or wetlands. All of this proposal except the

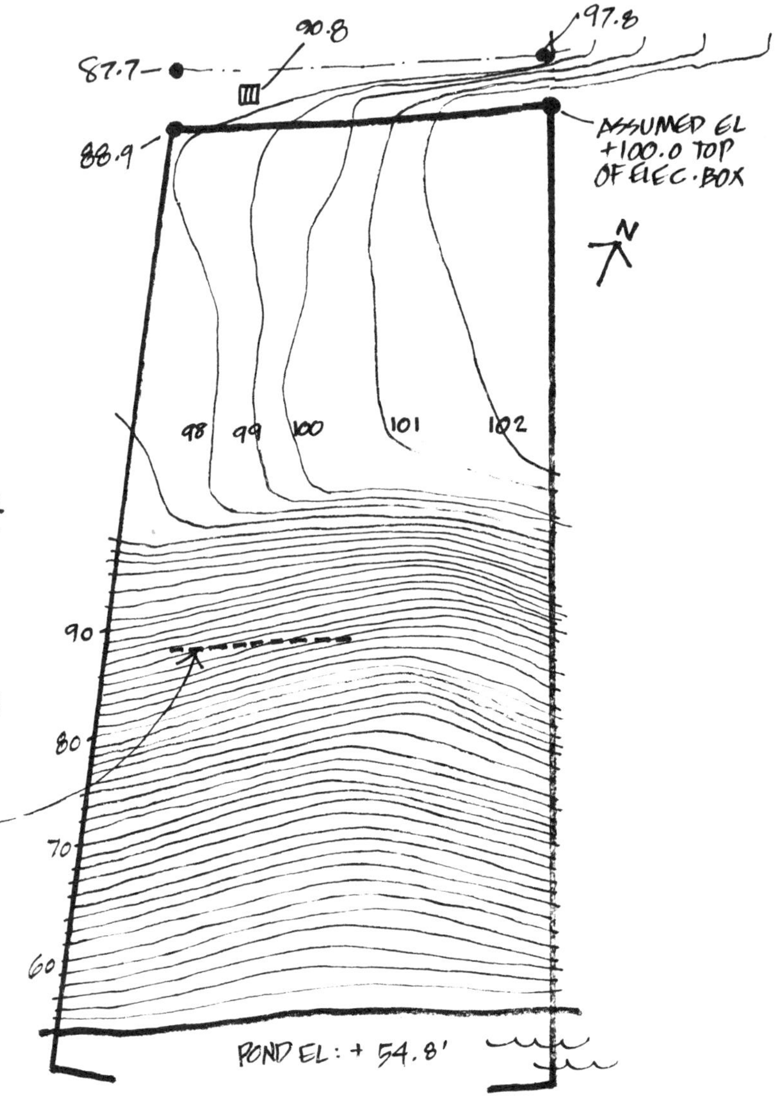

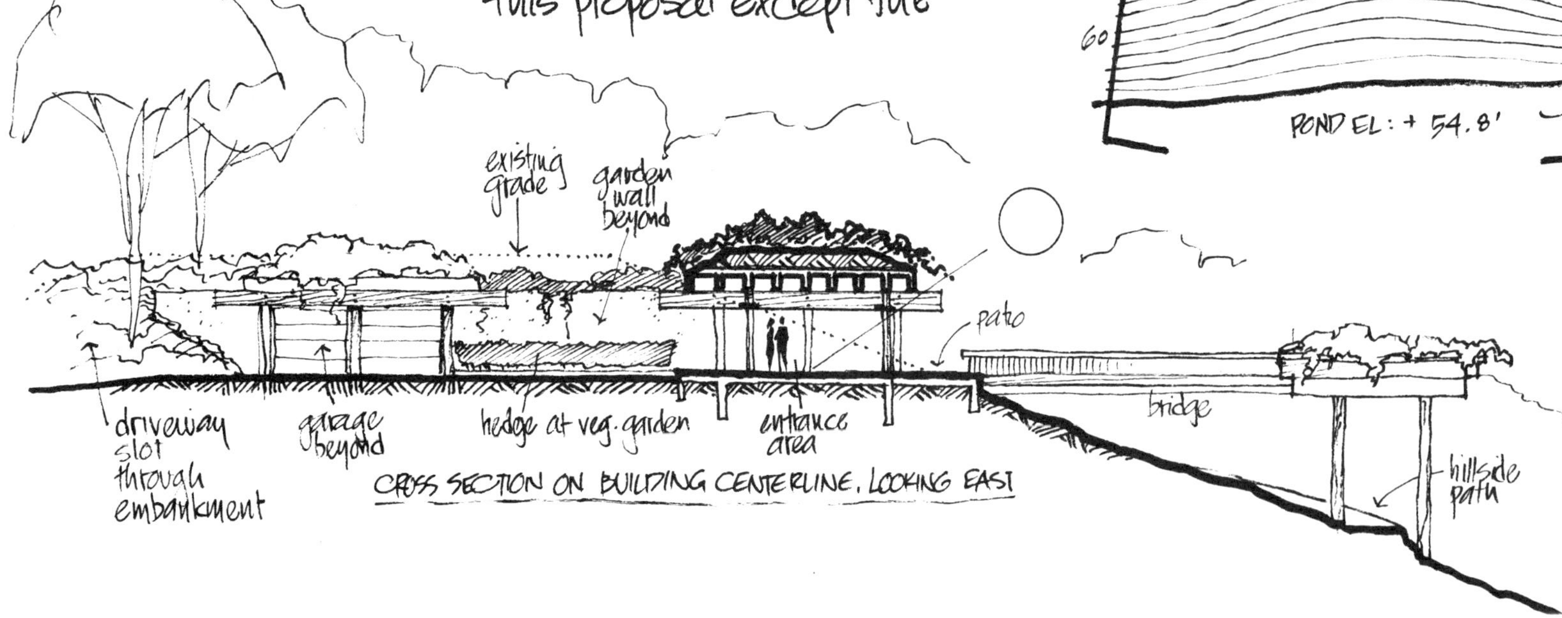

CROSS SECTION ON BUILDING CENTERLINE, LOOKING EAST

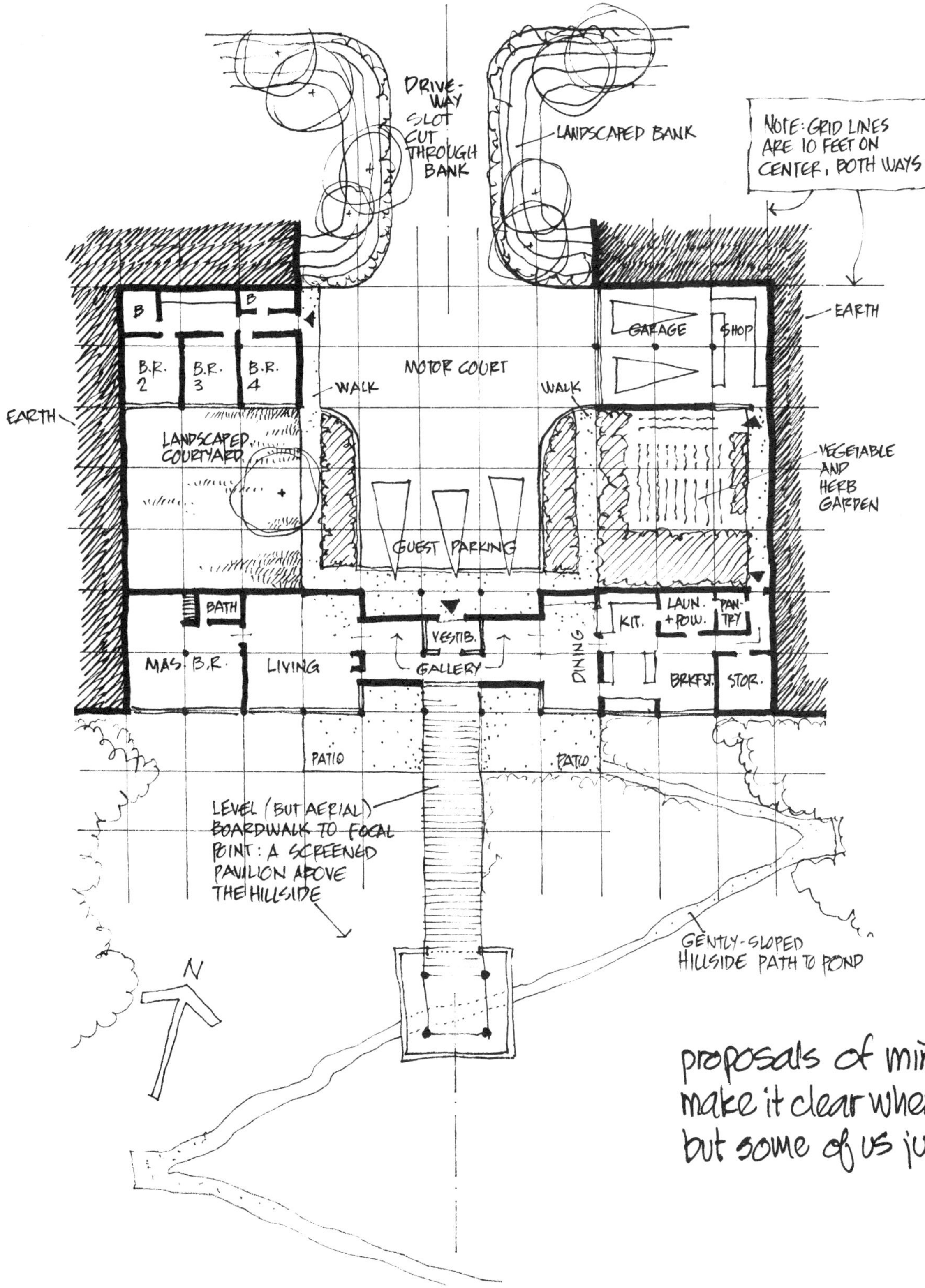

hillside path being more than 100 feet from the pond, it seemed likely to be approved.

We'll never know, however. The owners were never heard from again. My theory is that many people, perhaps having been stung by whopping bills from other professionals, go into hiding when they see these often-no-obligation proposals of mine in the mail. I make it clear when there's no charge but some of us just aren't responsive.

Excuse me, but isn't this where we came in?

Could be. It *is* getting to look awfully familiar, isn't it? But if you are to be convinced of the need for a new gentle architecture you have to become so mesmerized by it that the idea of building into the earth never seems foreign to you again.

The key words: silent, bright, dry, permanent, safe, ever-changing, and potentially beautiful.

Got it?

<u>This</u> potential client did remember to say thank you, and then some, even though nothing ever came of the proposal. Badly-needed low-income housing for the town of Orleans, here on Cape Cod, it simply ran into too many bureaucratic snags.

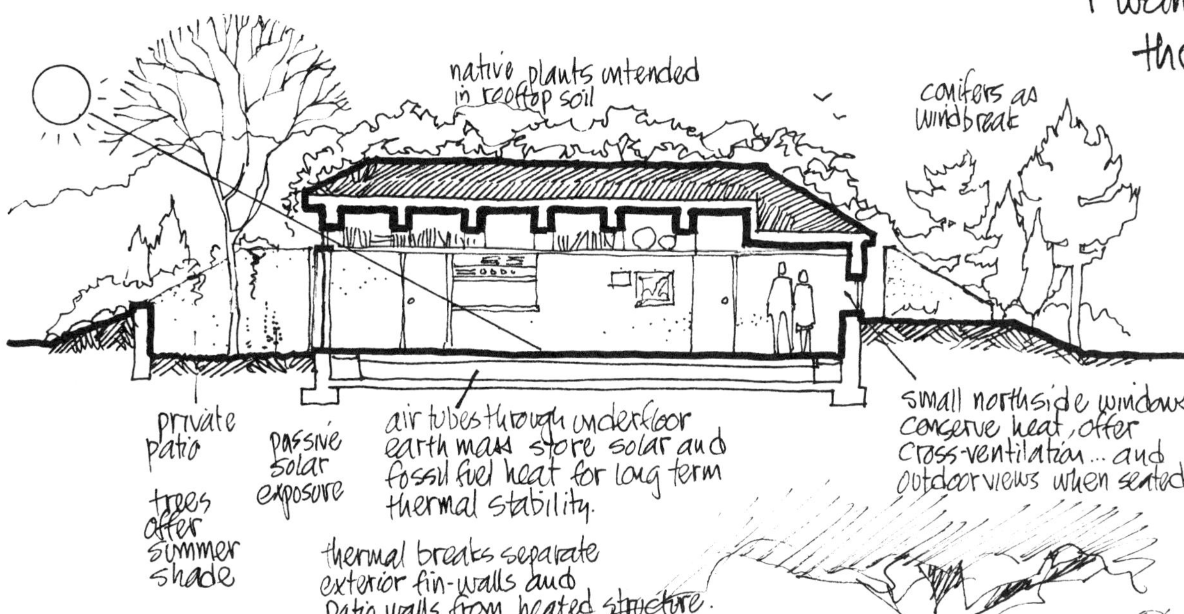

I want you to notice the high shelf above the sliding glass doors on the south side, just below the straight sunbeam line. It's a feature you'll see more and more in my work.

Run throughout a building, in the form of wall bands or shelves, this door-head-height theme unifies it all. see pages 112, 113.

private patios line the south facade of this six-unit underground housing complex.

Department of miscellany; another of those future cities with interchangeable, plug-in room-parts beneath permanent earth shelves; underground towers inspired by the purity of stacked porous paving blocks; and another entrance to something or other...

...and here: a Japanese/Cape Cod/Wrightian earth shelter, followed by solar village pyramids, the last bits of straight-line architecture before we go off into a rather extended period of circular designs.

HOUSE | PICTURE FOR JOHN, BOBBIE, & CO.

Wells

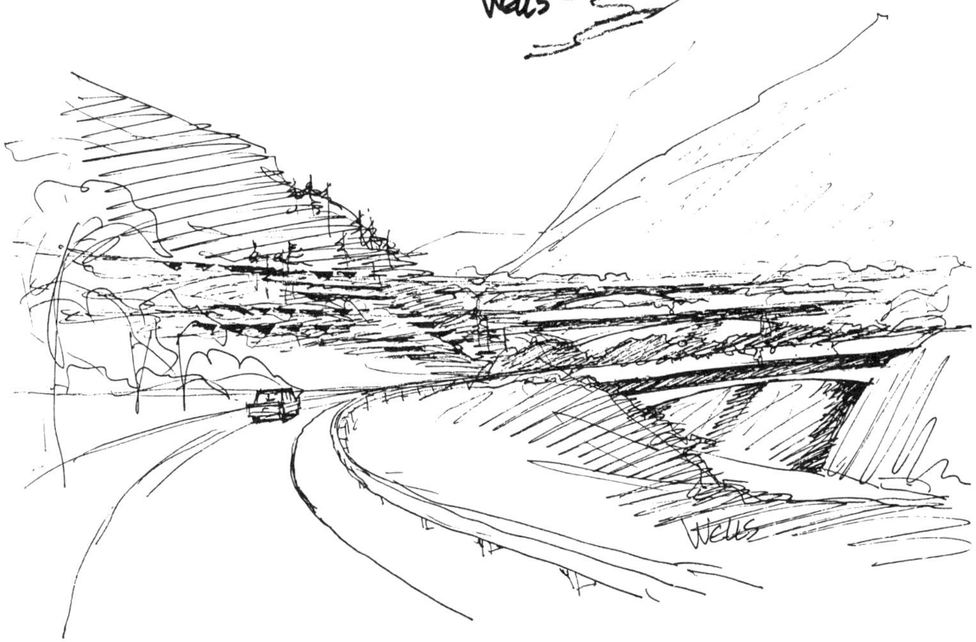
Wells

Cherry Hill, New Jersey, 1984. I was sure the developer who called on me for new design ideas for townhouses would be ripe for this idea. It shows what I know about the ripeness of people, particularly developers.

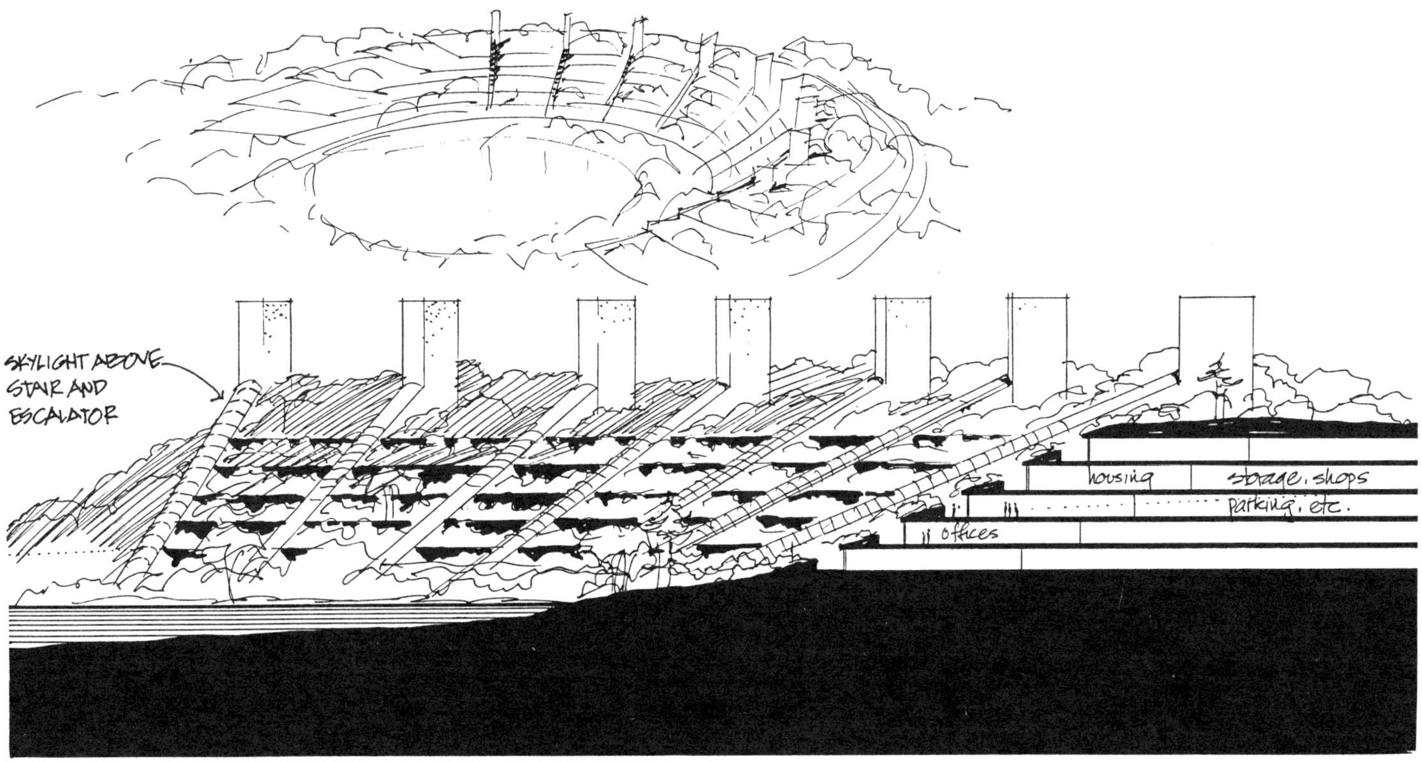

It's very flattering to be called all the way across America, out to the hills of Los Angeles, to visit the site of a proposed house. It beats a warm shower and a backrub,

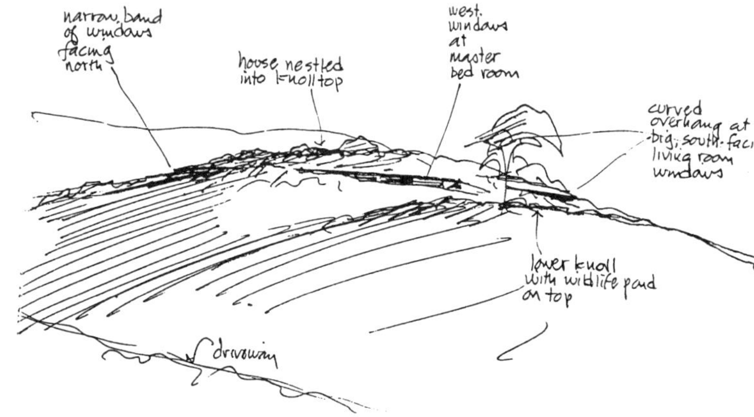

especially if the site is an entire hilltop, way out in the country, covered with acres of orange poppies.

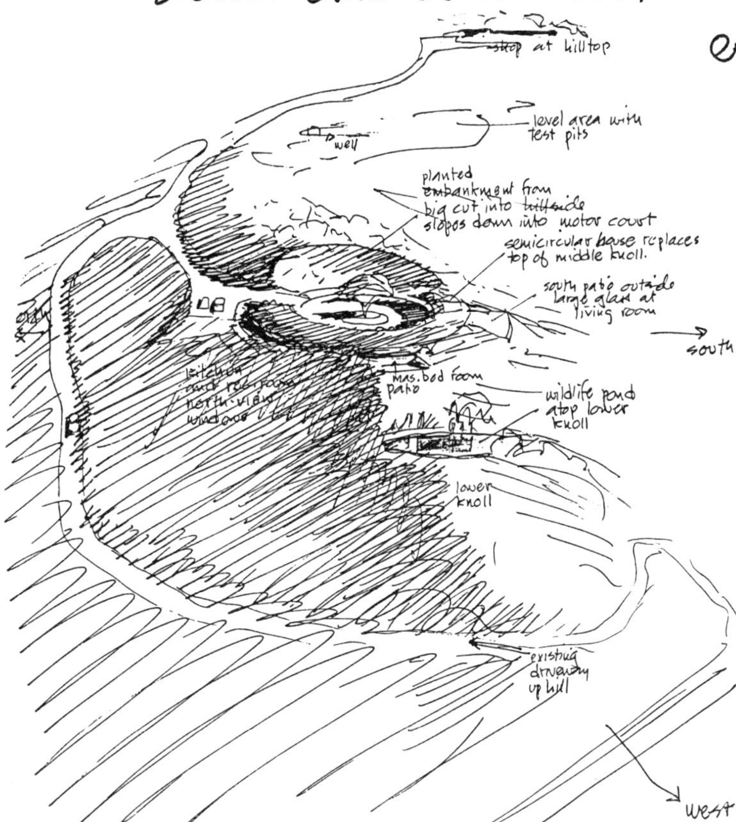

AERIAL PERSPECTIVE TAKEN FROM A POINT DIRECTLY ABOVE THE HIGHWAY EXIT.

I spent a long weekend in the area, walking the site, scribbling notes, and talking with the owners. The longer I looked at that round hill the rounder my ideas became until out burst this ring of rooms nestled so low into the crest as to seem a part of it.

Farther up the hill, the owners wanted a small workshop building (p.171). You can just catch a glimpse of it at left.

No, this one never got built, either. The owners suffered some sort of financial setback and had to give it up. While it lasted, our relationship was delightful, full of good idea-exchanges and open-mindedness.

I think I'll let you wander through the house by yourself for a while, and I'll meet you when you get up to the workshop.

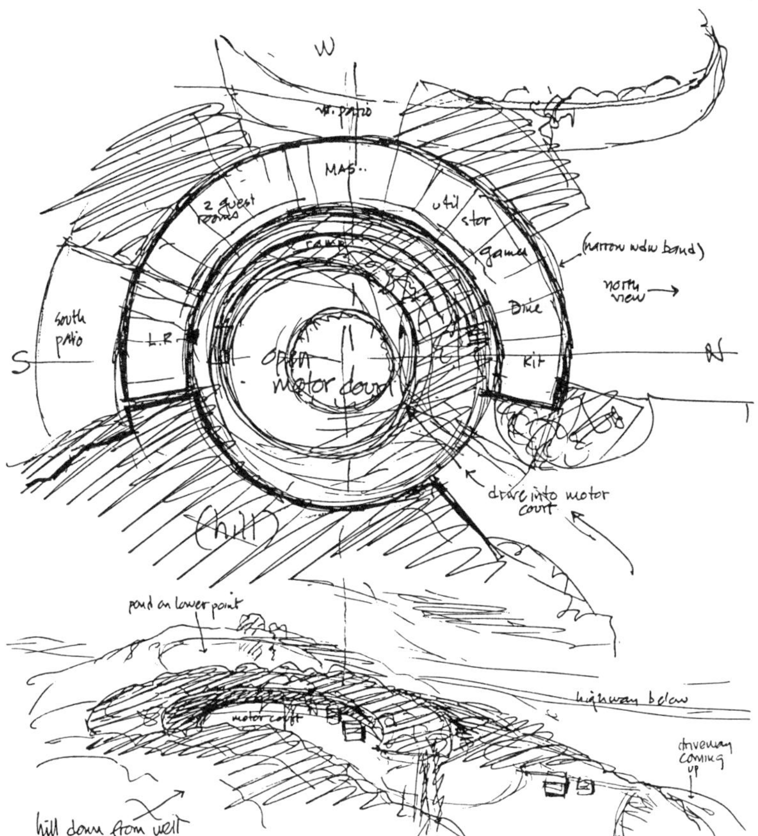

▸ VIEW FROM THE TOP OF A LADDER ATOP THE WELL HOUSE

S.W. side.

North is off to the right, south to the left.
The hill continues to climb from this rounded crest eastward (toward the bottom of the page).

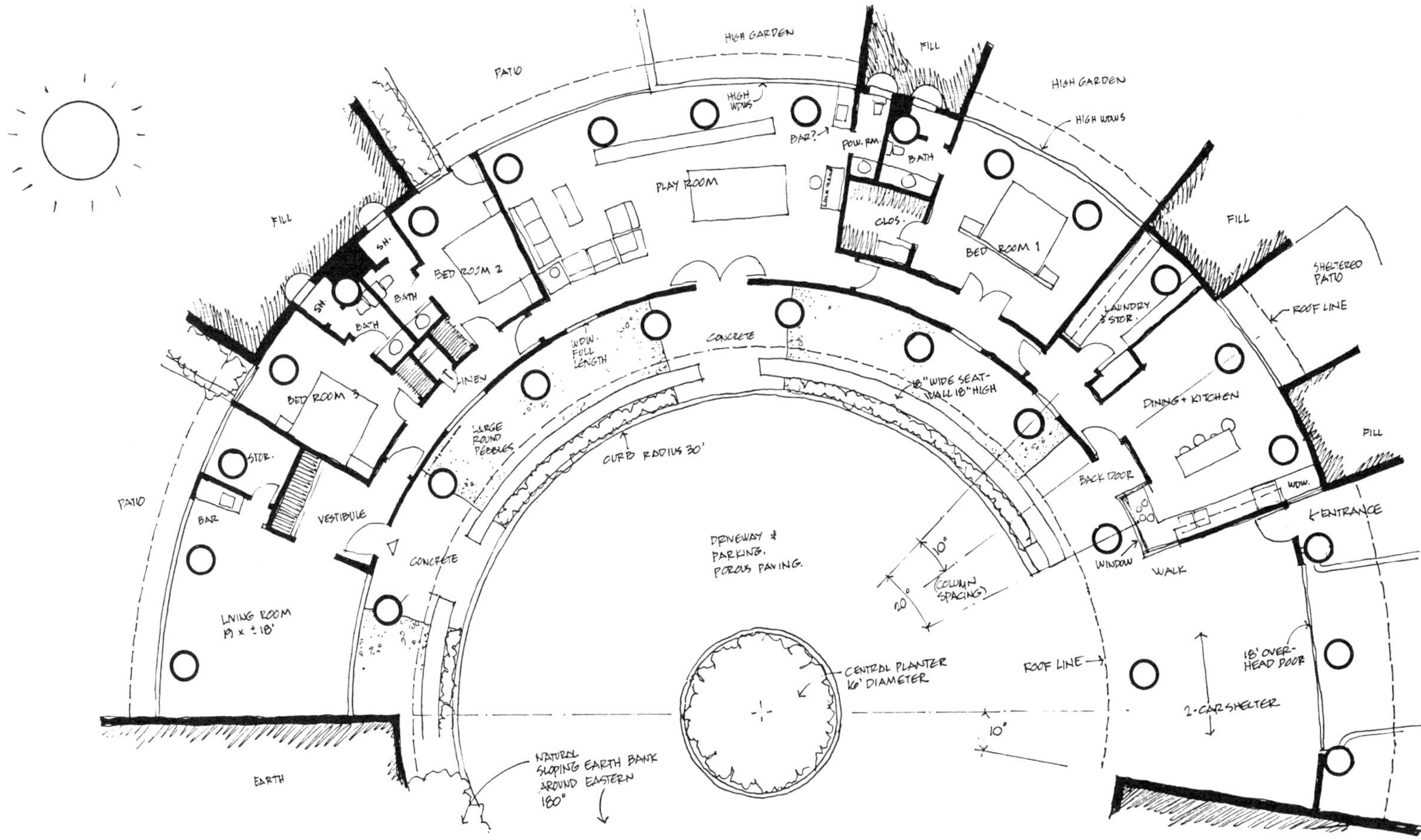

PARTIAL SECTION THRU COURTYARD, RAMP, AND STRUCTURE

SECTION SHOWING HOUSE SET DOWN TO LEVEL OF EXISTING GROUND SURFACE SO CONTOUR OF HILL (— — —) IS MAINTAINED

LIVING ROOM PERSPECTIVE

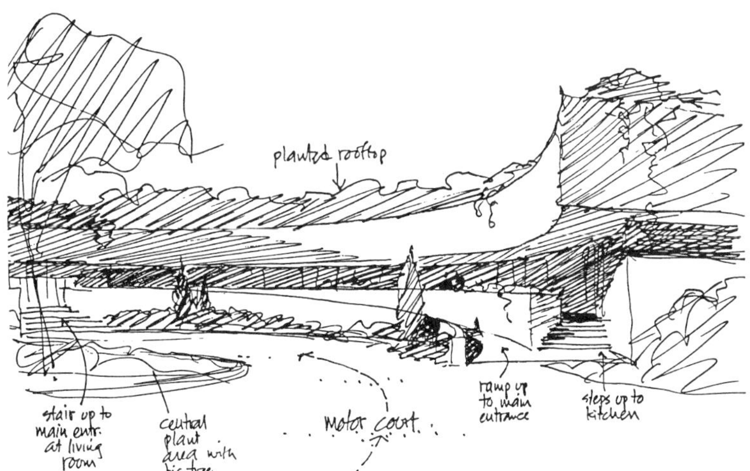

COURTYARD VIEW LOOKING SOUTHWEST

The lower knoll, or crest, obscures part of the freeway (below), and the hill climbs on up toward us as we look down from up near the workshop. ↓

↑ Arriving at the east side. Freeway in the distance, to the west.

The workshop is straight because long, straight things were to be made in it, and because the hillside was straight at that point. The southside solar area can be seen off to the right in this view looking from the west.

Next: two more circular designs before we get back into straight lines again.

172 A geodesic dome and a 4-level underground building encircling a pond, right in the heart of the city. If all the new buildings downtown were required to put life values first, the wildlife would stage a dramatic comeback.

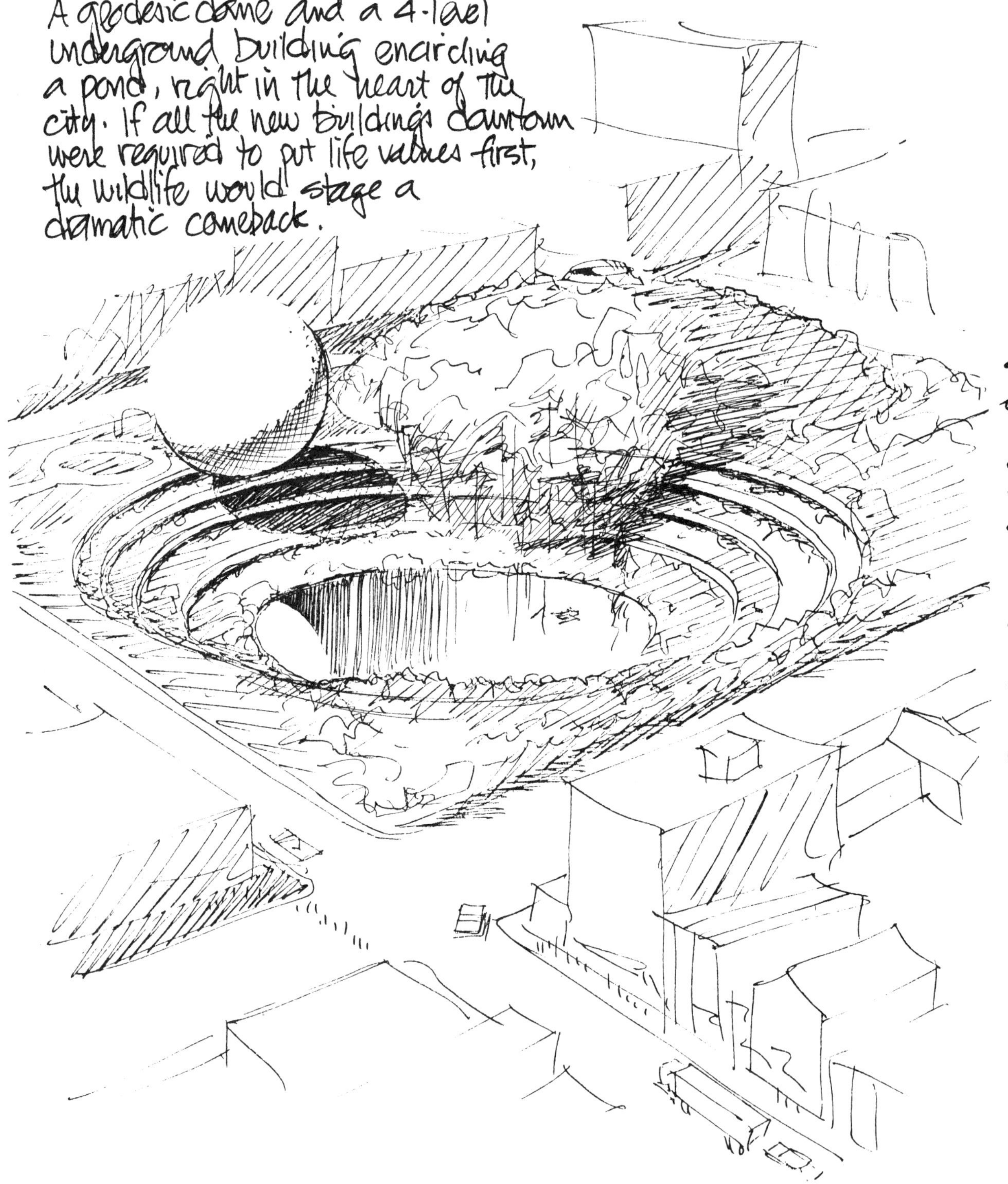

Ideal landscape budget: $ zero. Spread composted organic wastes, cover with leaves collected from the autumn streets, and voila! First: wildflowers, then shrubs, mosses, and vines, and then trees... without any help from us at all.

I was invited to teach an environmental course at Harvard in '83-84. Eight students signed up, and I soon found that they knew a lot more about everything than I did. After spending all those years in high school, undergraduate school, and graduate school they'd become a different species of animal to me.

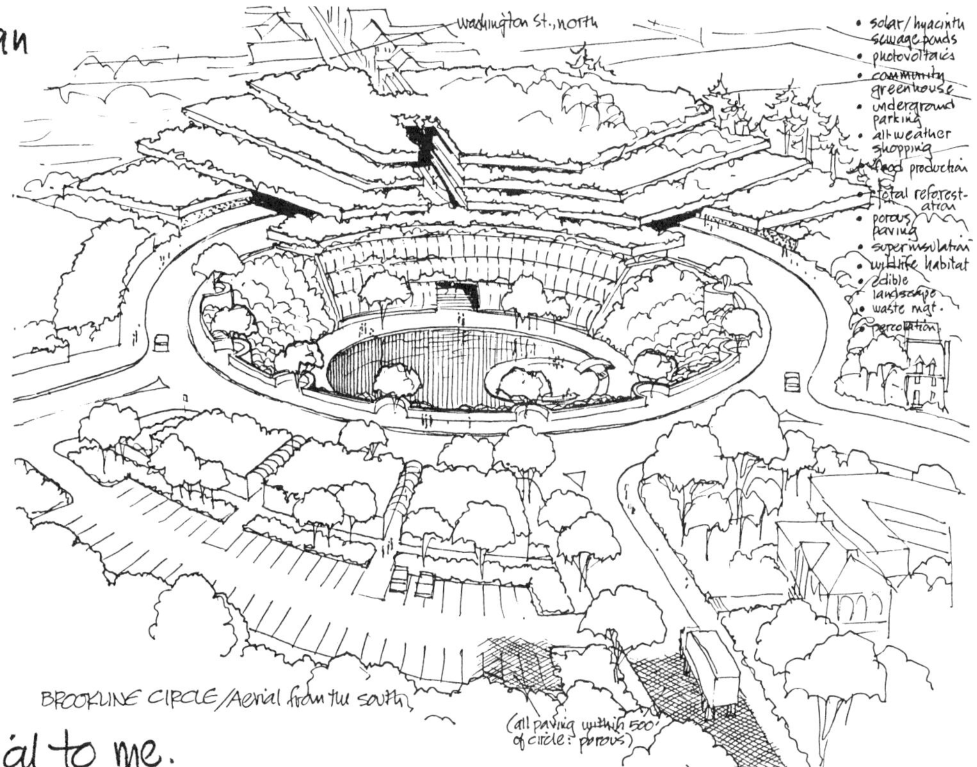

BROOKLINE CIRCLE/Aerial from the south

We worked out of different sides of our brains. or something. In any case, I was what is known as a flop. One of our projects was to redesign a rather ordinary intersection. I did the predictable thing (above); the students produced some esoteric mumbo-jumbo and almost no drawings. I couldn't *talk* architecture. They could.

↖ At one point we almost organized a summer tour of underground buildings in Sweden.

I've been tempted to name names in this book. After all, a lot of people put their faith, and sometimes their money, in me. I just don't want to embarrass anyone who couldn't, or wouldn't, go ahead with a project.

This couple in Pennsylvania went deeply into this project before deciding that with their kids about to go off to college there was no big need to leave the house in which they were already comfortably settled.

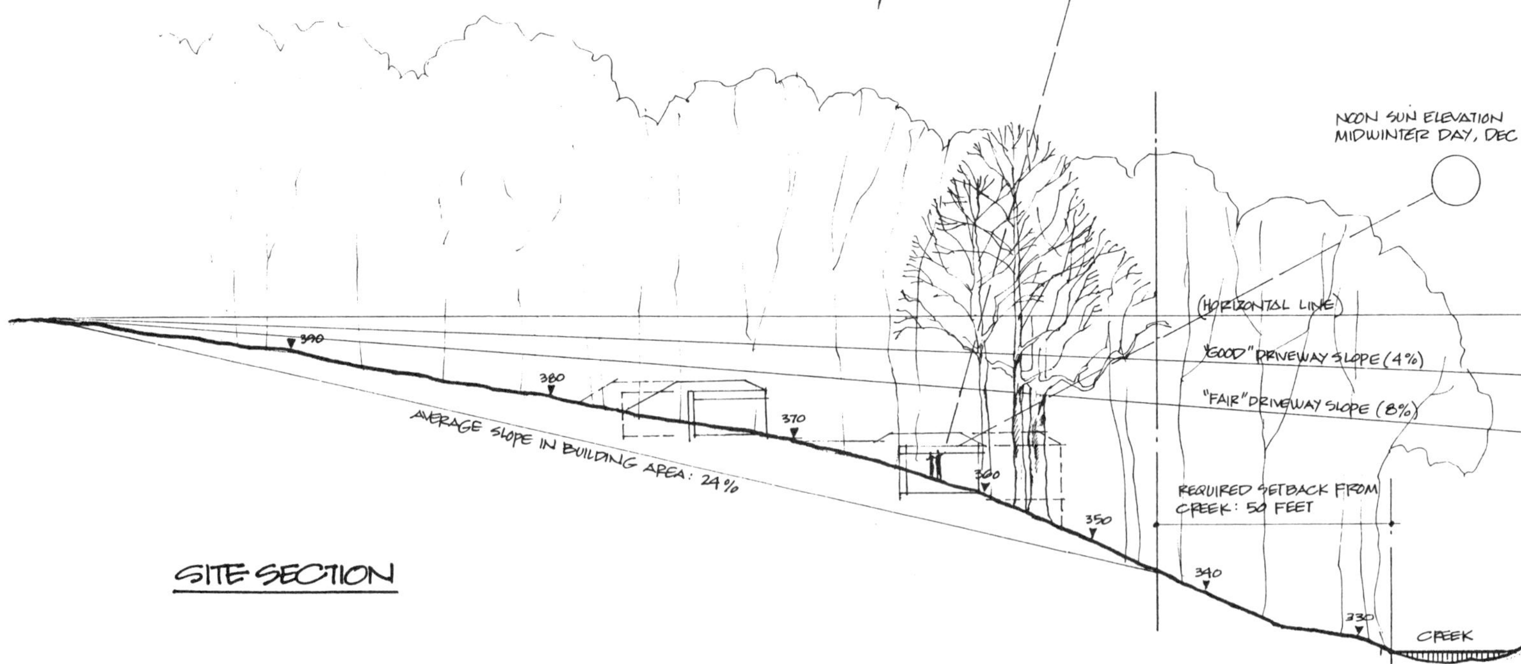

SITE SECTION

In order not to descend too steeply the driveway was made to come down around the front of the hillside, and into a slot between the buildings...

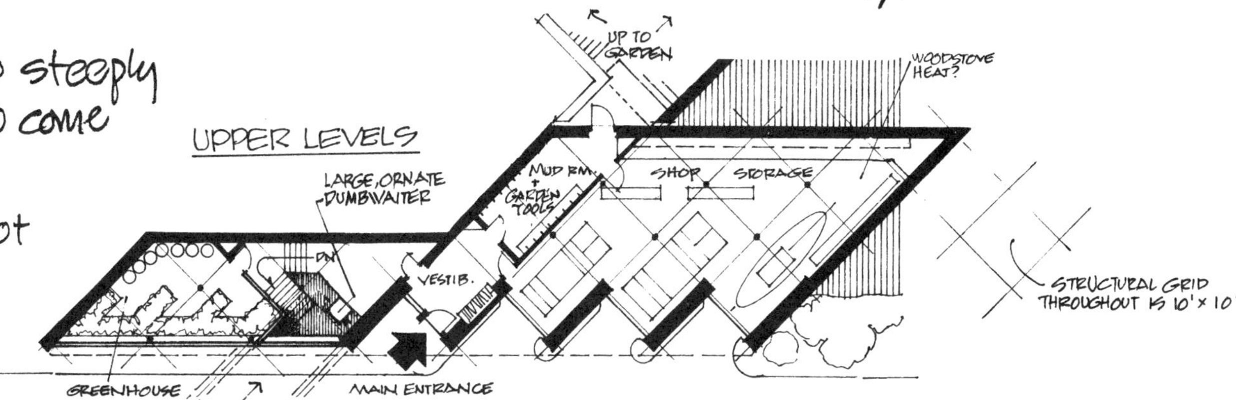

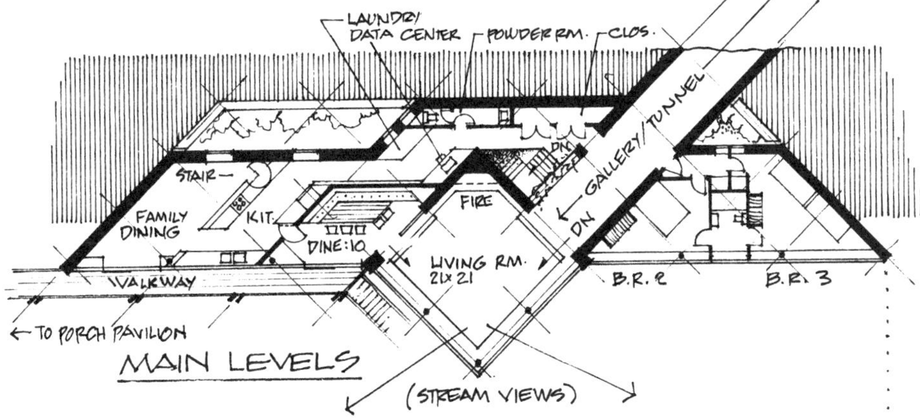

Notice that the entrance is in the garage building, above. From there, a stair (around a dumbwaiter for handling bundles) led down to the living level, and then yet another stair led further down to the lowest level still some 30 feet in elevation above the level of the creek.

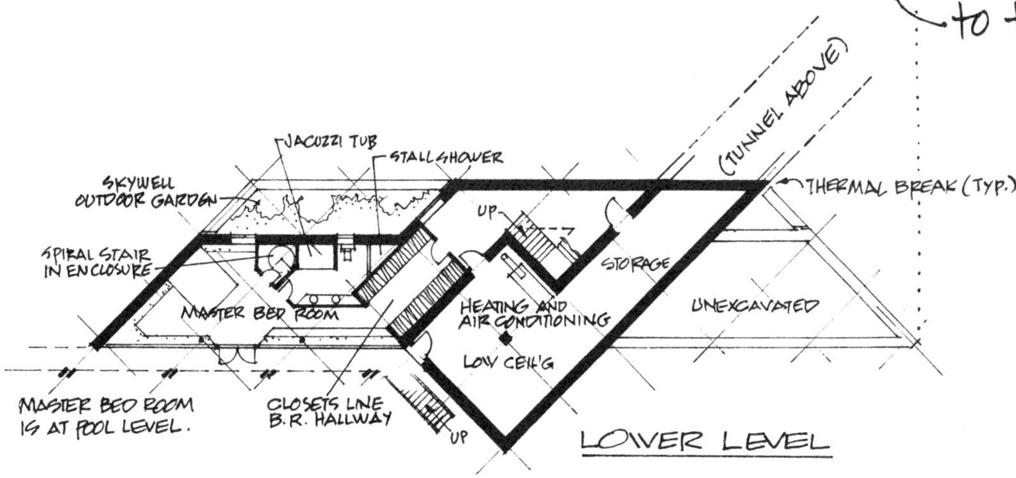

And what a creek that is! Lined with huge old sycamores and dotted with boulders, it flows through a silent forest of old trees.

This somewhat later design, like its partner to the left, eliminated the screened pavilion, added a swimming pool, and added an entrance directly into the house.

SOUTH ELEVATION / UPPER BUILDING

I like this original scheme better.

SOUTH ELEVATION / LOWER BUILDING

This is one client I've got to identify.

Irvin Cohen has been such a good friend and client for so many years he's got to take a bow whether he wants to or not. He and his wife, Lois, let me design their house, back in 1963. Since then, he's had me design his office and factory at Wyomissing, Pennsylvania, and he's been willing to let me use many innovations in the additions he's built there. Solar heat, task lighting, earth cover, natural landscaping, percolation beds, exterior insulation ... Here's another of those rare clients about which architects

dream. Mr. Cohen's latest addition to the Construction Fasteners complex was this earth covered parking lot entrance.

And now: a practice page....

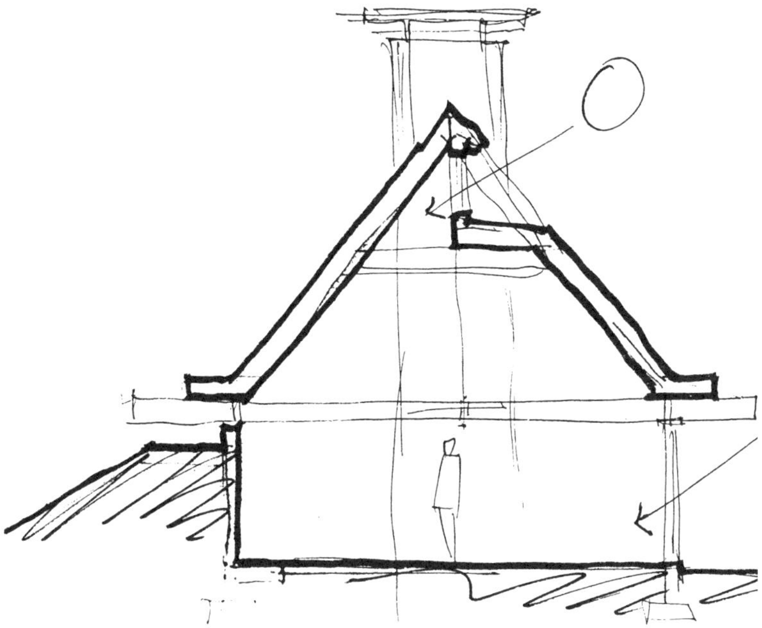

I shouldn't even think about big monster houses like this, especially for a single family, but I do love those long sweeping lines. Someday I'd like to do an earth-ridge house that looks like this. Would it live? Maybe it would if roots could be encouraged to grow thickly and survive during long dry periods.

Wouldn't it look great?

Refrigerating or freezing food is a big business today. (At Kansas City, abandoned mines that extend into hillsides, level, are easily kept at cold temperatures because of the lack of summertime warmth reaching them.)

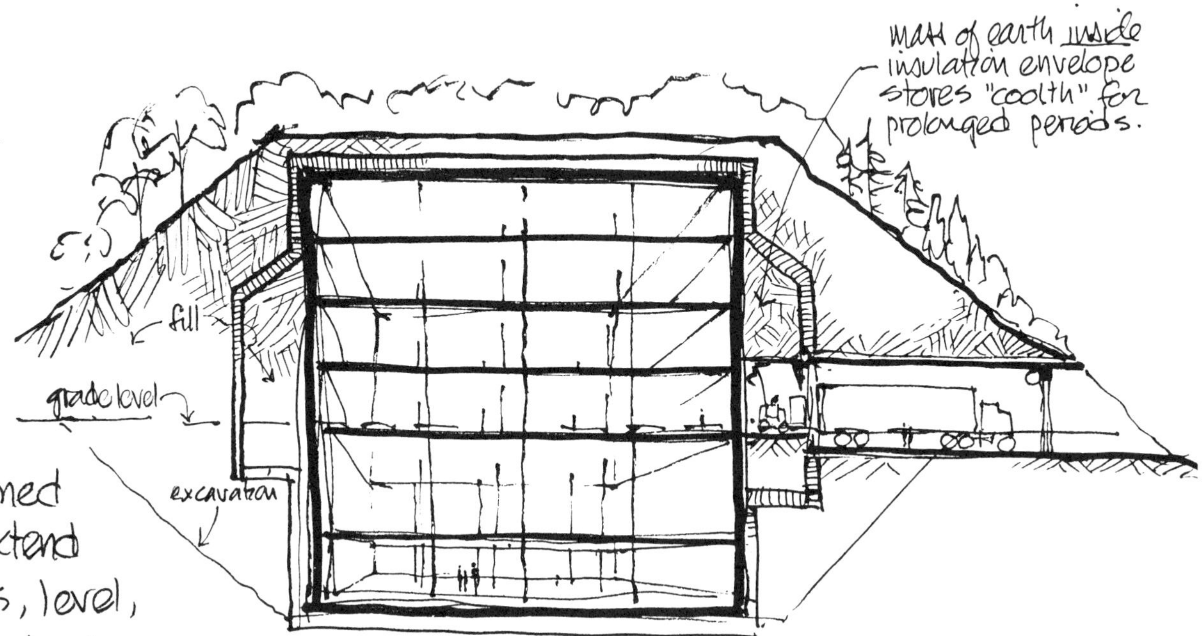

"THE ICE CUBE"

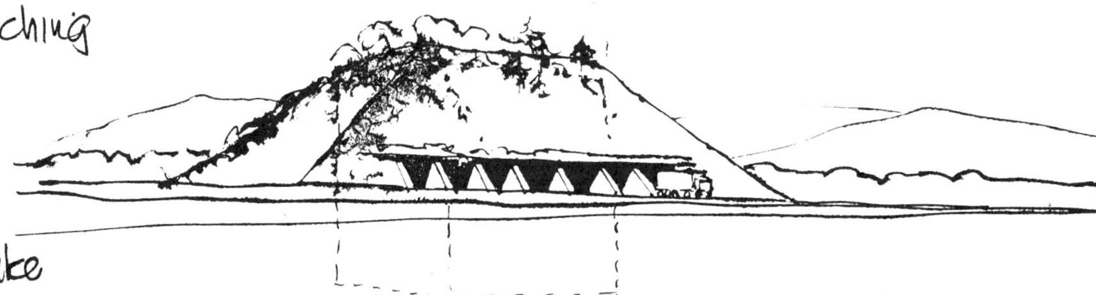

This was a recent attempt to take similar advantage of the earth's blessings. I couldn't find anyone with the direct experience and the credentials to advise me on the idea and again the momentum was lost with the client. It may still happen, though.

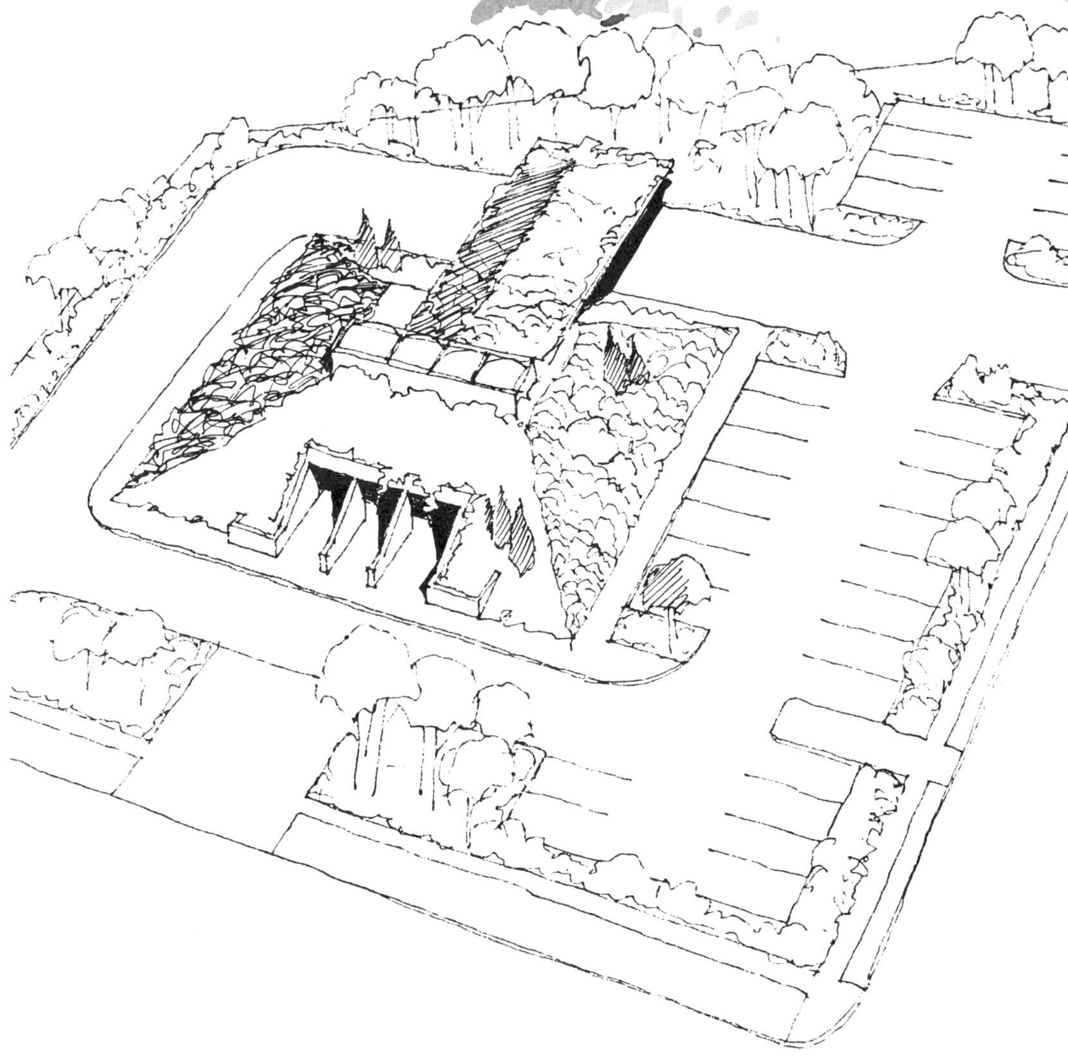

After hearing the employees complain about the discomfort of working in a local branch of a big-time bank I sent this to its headquarters. Now five years have passed. Should I expect an answer soon? Too bad I can't show you that original watercolor.

More miscellany ... a proposal for a new → church at Muncie, Pennsylvania, 1983; a proposal for a new parish hall at a Catholic Church here on Cape Cod; and the entrance to a house in Arkansas.

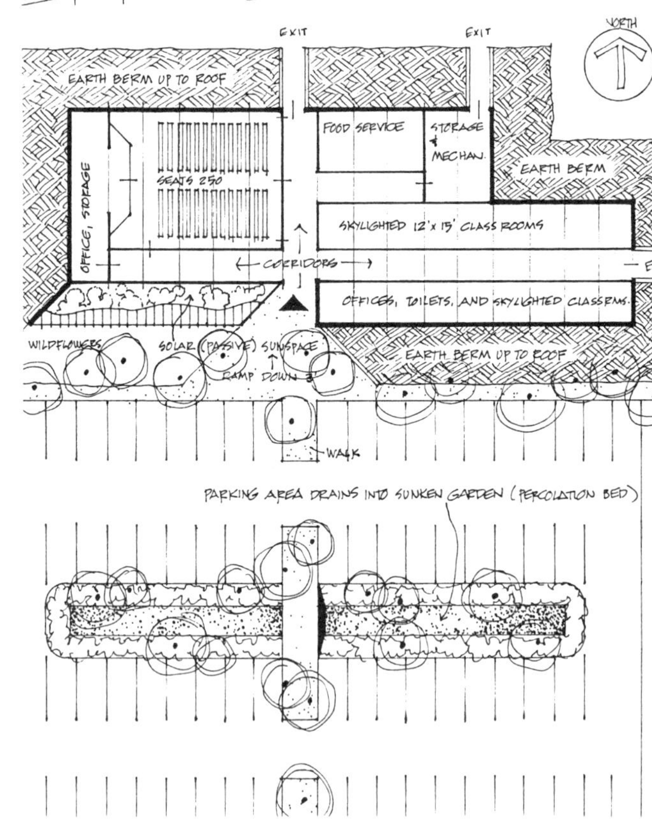

This is the only one of the three going ahead.

Do I see seagulls flying <u>below</u> an underground house? How I wish I could! These people put another damned Cape Cod cottage on that beautiful site.

This poolside house isn't dead. Just dormant. Why, it was only 18 months ago that I got that gracious "we'll be back" note from the owners.

April, 1989: Popular Science's editor was again kind enough to publish 2 designs of mine.... two designs for the same house! By rearranging the floor plans and the entrances a bit, I was able to adapt this by-now-familiar-to-you plan to work either way, making it suitable for almost any building lot big enough to accept it.

Northside entrance version

I sell construction plans, details, and specifications for either house (you must specify which one) for $100. In case you've forgotten, I am Malcolm Wells, Box 1149 Brewster, MA 02631.

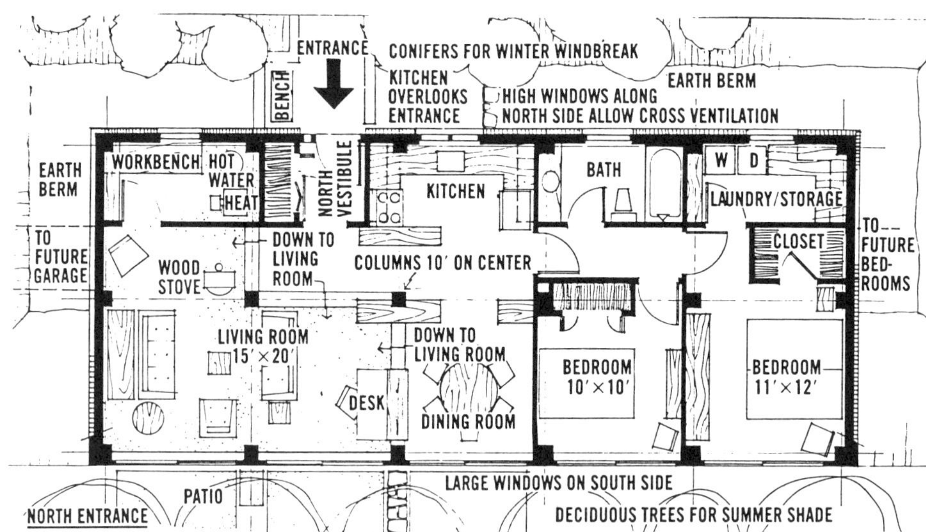

The perspectives look much better in color, of course. I'll try to remember to include them on the color pages.

Southside entrance version.

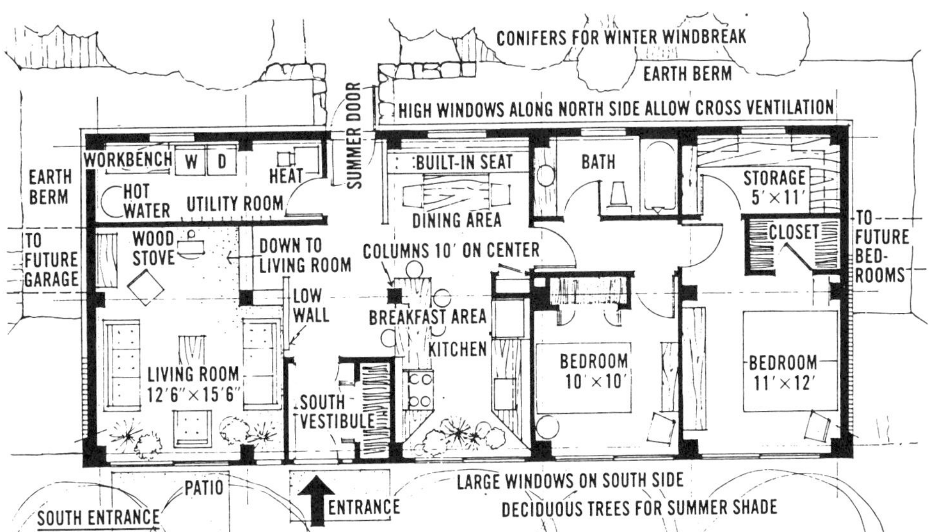

Notice that the house can be expanded very easily to either side... storage, bed rooms, garage, whatever.

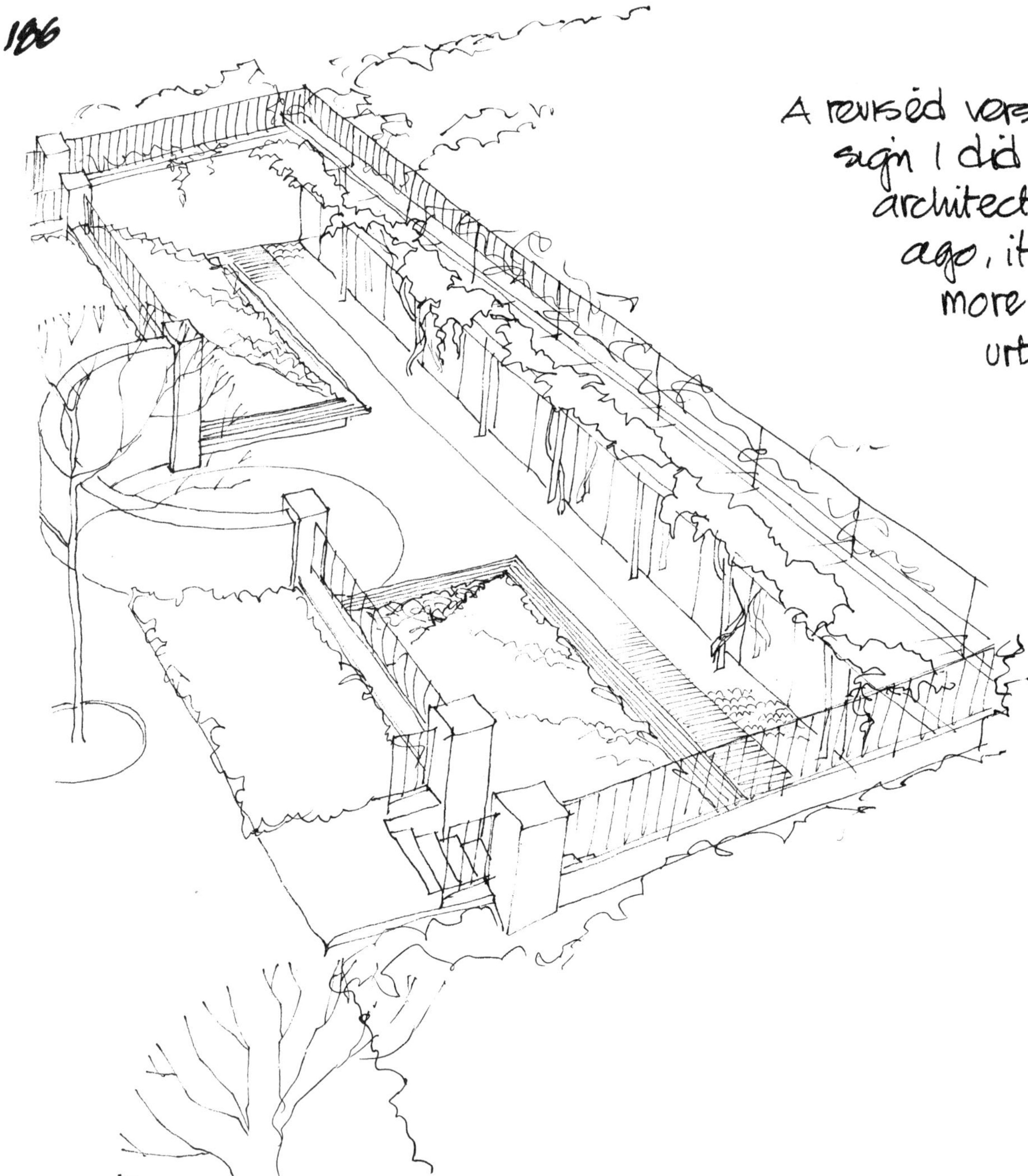

A revised version of a design I did for another architect a few years ago, it illustrates a more formal, more urban version of the same green idea.

The joint at the top of the wall in a buried building is always a problem. So many ~~many~~ materials come together there it's almost impossible to hope that all the water will be kept out. Sometimes I think the best solution is to raise the entire roof above the berm and let it act as a safety rail and drip ~~place~~ edge.

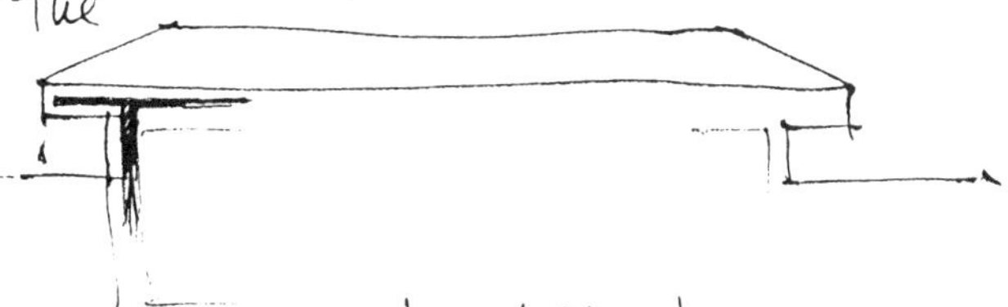

But then the roof overhang, if it is not to be a severe heat bleeder, must be so thick it might as well not be an overhang at all. That leads to this approach: big and klunky. All you can say for it is that it stays out of trouble, and, if leaks develop there, you can walk right up to them and fix them.

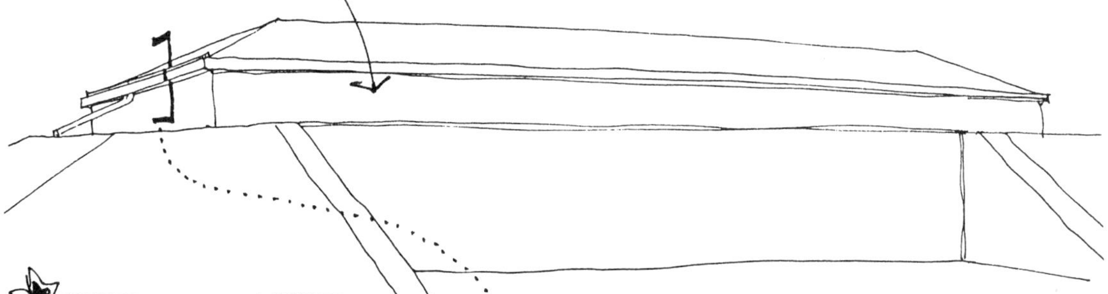

A page from one of my journals. I'll keep trying to find a way to insulate the cantilevered edge of an earth-covered building until I can prove that it is possible.

But wouldn't it be nice to bring the rooftop earth to a knife edge, as we (sort of) did

at the Gallery? It would look as if the roof garden just happened to lift itself up for a while to have a look around.

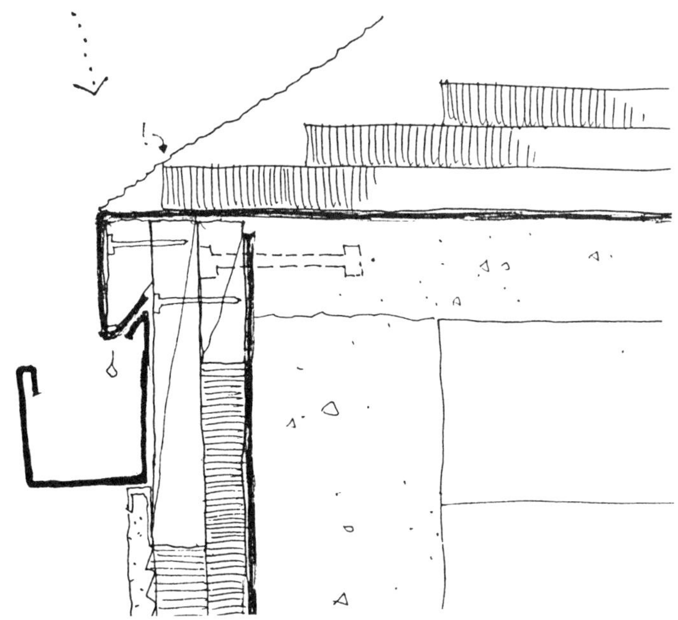

My most recent book is called Classic Architectural Birdhouses and Feeders*(CABAF, here in the office). Among its offerings is this "underground" birdhouse, high up on a pole. I built such a birdhouse, and was happy to see purple finches building a nest in it. The big haystack-shaped mound of grass thrived for a while but the drought caught up with it and by August the mound was golden, not green. That original mound had very little absorbtive material inside. That's what led to the development of this water-holding version. It's bound to work a lot better but I haven't had a dry season yet in which to test it.

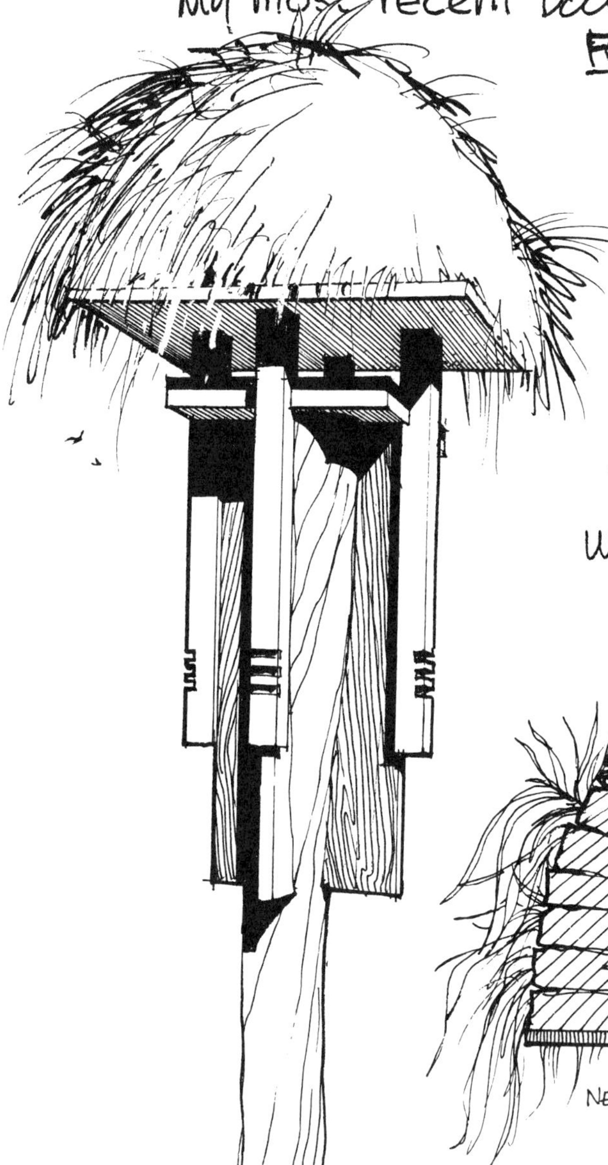

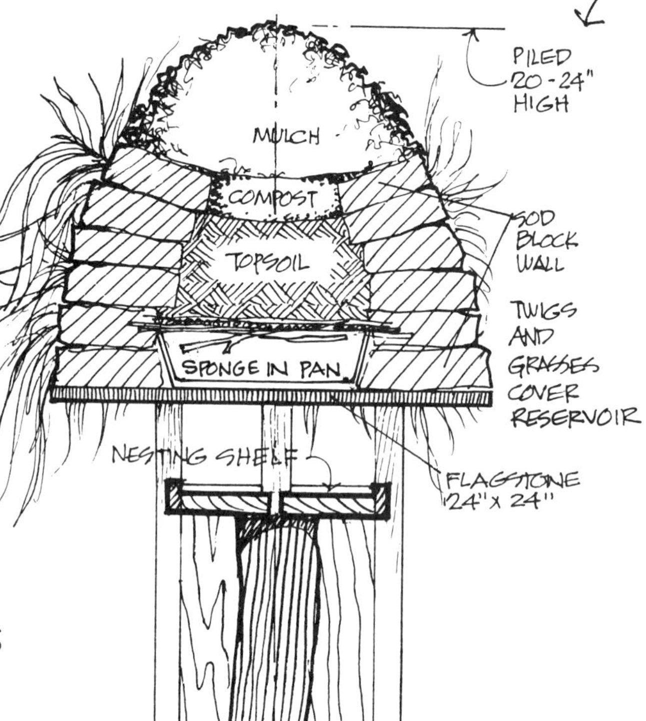

For a copy of this magnificent book, send me a mere $9.95 for a postpaid copy.

Do you still remember who I am, or must I tell one more time?

*92 pp. 8½×11" Softcover. Color photographs. Plans for 26 structures.

Almost 20 years ago, a student I was never able to identify made this poster for my Cranbrook talk. Since then, I've used the upside-down-houses theme many times, most often in my Underground America Day notices. U.A. Day, by the way, is always on May 14th. Don't forget!

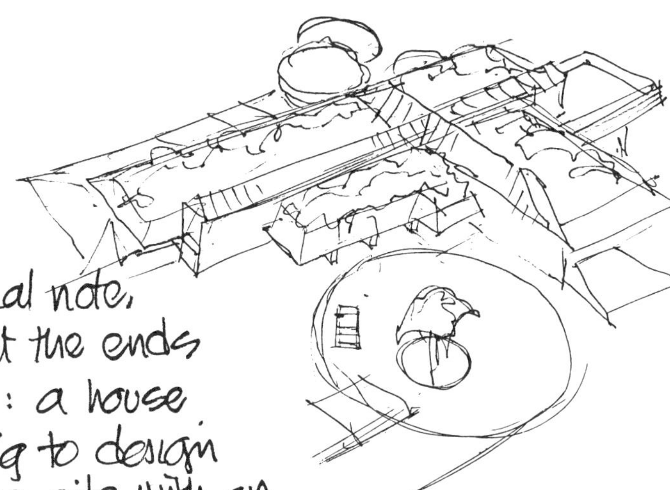

And now a final note, as they say at the ends of newscasts: a house I'm just starting to design for a ridge-top site with an ocean view in northern California. The ridge, like that of the house on p. 94, runs north and south.

The End.

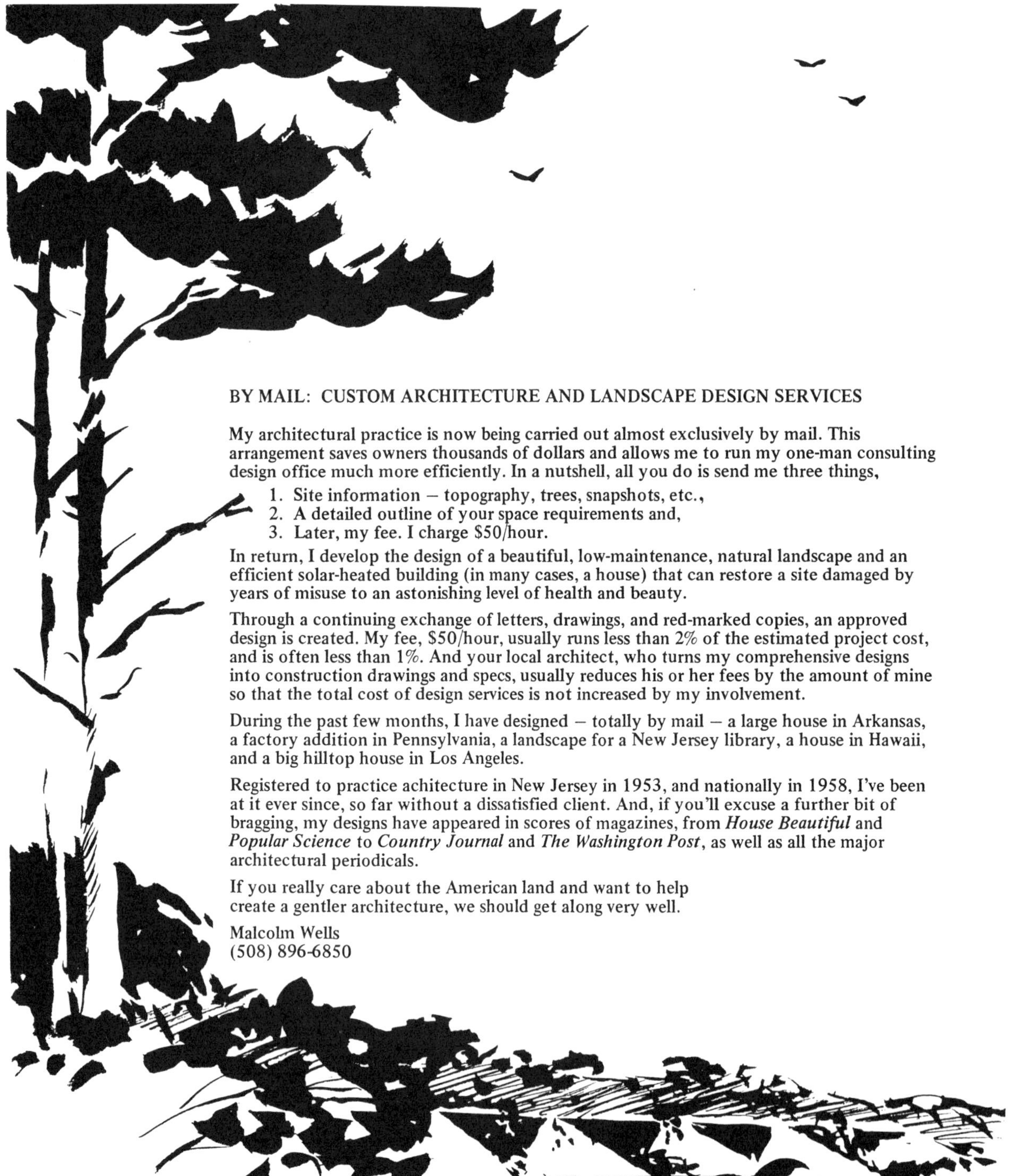

BY MAIL: CUSTOM ARCHITECTURE AND LANDSCAPE DESIGN SERVICES

My architectural practice is now being carried out almost exclusively by mail. This arrangement saves owners thousands of dollars and allows me to run my one-man consulting design office much more efficiently. In a nutshell, all you do is send me three things,

1. Site information – topography, trees, snapshots, etc.,
2. A detailed outline of your space requirements and,
3. Later, my fee. I charge $50/hour.

In return, I develop the design of a beautiful, low-maintenance, natural landscape and an efficient solar-heated building (in many cases, a house) that can restore a site damaged by years of misuse to an astonishing level of health and beauty.

Through a continuing exchange of letters, drawings, and red-marked copies, an approved design is created. My fee, $50/hour, usually runs less than 2% of the estimated project cost, and is often less than 1%. And your local architect, who turns my comprehensive designs into construction drawings and specs, usually reduces his or her fees by the amount of mine so that the total cost of design services is not increased by my involvement.

During the past few months, I have designed – totally by mail – a large house in Arkansas, a factory addition in Pennsylvania, a landscape for a New Jersey library, a house in Hawaii, and a big hilltop house in Los Angeles.

Registered to practice achitecture in New Jersey in 1953, and nationally in 1958, I've been at it ever since, so far without a dissatisfied client. And, if you'll excuse a further bit of bragging, my designs have appeared in scores of magazines, from *House Beautiful* and *Popular Science* to *Country Journal* and *The Washington Post*, as well as all the major architectural periodicals.

If you really care about the American land and want to help create a gentler architecture, we should get along very well.

Malcolm Wells
(508) 896-6850

$7 postpaid.
My old stand-by. Eleven years old and still going strong. Now in its 8th printing, this widely-reviewed book has sold over 90,000 copies! This 8½x11" paperback has 88 pages, 170 illustrations, and 17 house designs (not construction plans!). There are also businesses, institutions, and information on codes, zoning, soils, waterproofing, insulation, and landscaping.

UNDERGROUND DESIGNS

UNDERGROUND PLANS BOOK I

$15 postpaid.
22x11". 44 pages. 8 different large-scale floor plans, sections, details, etc. Recommended by *Popular Science*, *New York Times*, *Solar Age*. Easy-to-follow illustrations.

CLASSIC ARCHITECTURAL BIRDHOUSES AND FEEDERS

$9.95 postpaid. 8½x11". 92 pages. After building birdhouses for family and friends for over 40 years, I decided to offer the 26 designs in easy-to-follow, step-by-step construction plans. My hope was to get us away from the usual dumb box so commonly perceived as the final word on architecture for the birds. The book must have touched a nerve; it's now into its second 10,000-book printing and has drawn gratifying reviews from major magazines across the country.

". . . unique and welcome" (*Wild Bird*) ". . . lively, environmentally fitting" (*American Library Association*) ". . . clever, inconspicuous blend of aesthetics . . . and besides, how many opportunities do you get to commission an architect to design your birdhouse?" (*Mother Earth News*) ". . . a book to read, enjoy, and cherish" (*Express*)

". . . classy birdhouses . . . (*Minneapolis Star*) ". . . a fine gift suggestion" (*Boston Herald*) ". . . beautiful birdhouses" (*Houston Chronicle*)

Index

"...the earth-cast buildings
of Riviera architect Jacques Couelle..."
(See page 5.)

In 1971, The Today Show
invited me to talk about
underground buildings.
I had just four days in which to build this model, so I worked
non-stop. Rooftop wildlife ponds, organic gardens on each level,
composted wastes... it generated a flurry of letters, and then: nothing.

From "Penn's Sylvania", an article I wrote for Charrette,
the journal of the Pennsylvania Society of Architects, in 1968.

From "Nowhere to Go But Down", Progressive Architecture,
February, 1965. (see page 6.)

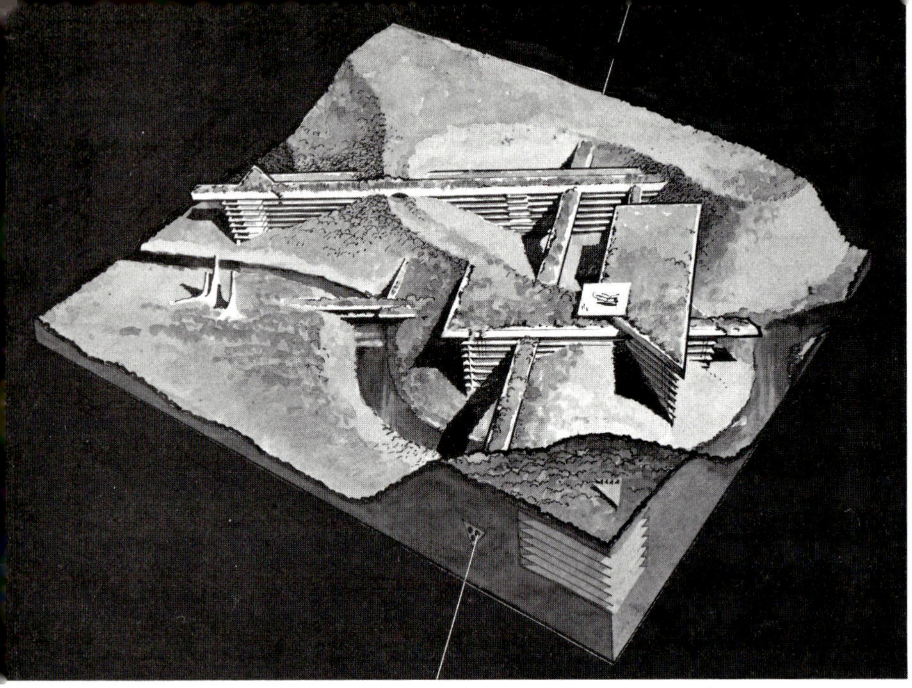

"A Town of 5000." Proposal, 1970.

1969. "Museum of The Future", done for the American Association of Museums, the design featured earth cover, solar heating, underground parking, on-site composting of organic wastes, a wildlife refuge, natural plantings, and a great percolation bed/retention basin for the site's run-off. (See page 14.)

"Low and Bermed", the little house featured in the June, 1984, issue of Popular Science. (See page 138.)

One of two designs possible with a single floor plan, this version has its main entrance on the north....

....and this, on the south. Both were published by Popular Science in April, 1989. (See page 184.)

1978. The newly-completed Plant Science Building of The New York Botanical Garden's Cary Arboretum at Millbrook, New York. Photo: Artog. (See page 46.)

Photo: Ken Basmajian.

Courtyard, Cherry Hill underground office.

My first underground building, the office at Cherry Hill, New Jersey. 1970. (See page 22.)

Photo: Artog.

The tireless champion of all that is good, seen here upstaging his own building. House at Cape Cod. 1980.
Photo: Lou Penfield.
(See page 96.)

Another view of it, featuring the gifted touch of landscape designer Shirley Dorkin of Cherry Hill, N.J.

Her design touch is equally in evidence in this interior shot of the Cape Cod house. Note the low-cost, low-maintenance floor of rough concrete, stained and sealed.

1978. The original design for the Cape Cod house, it would have sacrificed view for sunlight, and I decided against it.

Locust Hill at Raven Rocks, Beallsville, Ohio. Don't miss the story of this fascinating project. It starts on page 82.

A cross section through Locust Hill, showing all the ways in which sunlight will penetrate deep into this hill-sheltered structure. South is to the left, north to the right.

If the gardens I've seen at Raven Rocks are any indication, this one will be almost unbelievable. To me, it's as much a part of the design as the walls and roof, expressing as it does, so much about life, sunlight, earth, and the miracle of growth.

Great care is taken each step of the way at Locust Hill out of respect for the materials themselves and the jobs that will be asked of them over the coming centuries.

The day I got to pose as a construction man, vibrating the concrete as it went into the form at Locust Hill. With me, and doing the real work, were Warren Stetzel and Tim Starbuck.

The Sidwell house at Raven Rocks. (See page 88.)

The Underground Gallery at Brewster on Cape Cod, where I sit as I write these words. The building, still incomplete, cries out for the plants that will, in time, make it look almost as good as I pictured it when I did this rendering. (see page 108.)